Motorbooks International Illustrated Buyer's Guide Series

Illustrated DODGE & PLYMOUTH MUSCLE CAR BUYER'S ★ GUIDE™

Peter Sessler

Acknowledgments

I thank all those who contributed photos and information to this book with special thanks to Richard Dangler, Wayne Hartye, Galen Govier, Alan Linsky, Guy Morice, Alan Ralston, Don Rook, Don Snyder, and Chrysler Corporation. All photos by the author, except where indicated.

First published in 1995 by Motorbooks International Publishers & Wholesalers, PO Box 2, 729 Prospect Avenue, Osceola, WI 54020 USA

Library of Congress Cataloging-in-Publication Data
Sessler, Peter C.
Illustrated Dodge & Plymouth muscle car buyer's guide/Peter C. Sessler.
p. cm. —(Motorbooks International illustrated buyer's guide.)
Includes index.
ISBN 0-87938-975-3
1. Dodge automobile—Purchasing. 2. Plymouth automobile—Purchasing. 3. Muscle cars—Purchasing. I. Title. II. Title: Illustrated Dodge and Plymouth muscle car buyer's guide. III. Series.
TL215.D6S477 1995
629.222'0296—dc20 94-39838

On the front cover: With 440ci under the hood, Plymouth's upscale muscle offering for 1969, the GTX, gave nothing away to the cheaper and more popular Road Runner. This one belongs to Greg Rader of Lakeland, Florida. *Mike Mueller*

Printed and bound in the United States of America

Contents

	Introduction	**4**
	Investment Rating	**5**
Chapter 1	**1955–65 Chrysler 300 Letter Series**	**6**
Chapter 2	**1956–58 Plymouth Fury**	**40**
Chapter 3	**1956–60 DeSoto Adventurer**	**48**
Chapter 4	**1967–74 Plymouth Belvedere GTX and Road Runner**	**59**
Chapter 5	**1967–70 Dodge Coronet R/T and Super Bee**	**74**
Chapter 6	**1967–74 Dodge Charger and Super Bee**	**80**
Chapter 7	**1967–74 Plymouth Barracuda**	**91**
Chapter 8	**1970–74 Dodge Challenger**	**105**
Chapter 9	**1968–73 Dodge Dart GTS, Dart Sport, Demon 340, Plymouth Valiant Duster 340**	**113**
Chapter 10	**1970–71 Plymouth Sports Fury GT and S/23**	**118**
Appendices	**Production Figures**	**120**
	Clubs	**127**
	Index	**128**

Introduction

The image that comes to mind for many people when the subject of Chrysler automobiles is brought up is that of big, rumbling muscle cars. However, I think in this book you will discover that Chrysler automobiles have more depth and scope than this "muscle bound" reputation suggests. Chrysler has built some of the most fabulously unique cars with incredible styling and performance.

Chrysler, along with the rest of the auto industry, was in a very strong position after the war. It sold all the cars it could build to a car-starved public even though postwar styling was about as conservative as you could get. After 1949, the public showed a preference for the sleeker offerings from Ford and General Motors to the point that Chrysler fell to third place behind Ford in 1952. The situation had worsened by 1954, but then the Virgil Exner-styled cars of 1955 reversed Chrysler's fortunes. By 1957, Chrysler had clearly established itself as the U.S. auto industry's styling leader. The stunning 1957 Chrysler lineup shook the industry. Unfortunately, Chrysler was unable to follow-up, because, as Lee Iacocca has said, it's styling that sells cars, but its value that keeps customers coming back. This created a catch-22 situation because practical cars did not sell unless they were also attractive. The fifties and sixties were especially style-sensitive eras. It became very important not to be seen driving last year's model.

By the early 1960's Chrysler's styling had run amok. What had originally been clean, elegant, and innovative was taken to such extremes that the cars no longer appealed to the American public. During the rest of the decade Chrysler played catch-up, following rather than leading in terms of styling. They did, however, produce one of the 1960's most visually exciting automobiles, the 1968 Dodge Charger.

The relative prosperity of the 1960's gave way to the trauma of the 1970's—two oil supply shocks and near bankruptcy for Chrysler. But Chrysler again showed resiliency, bouncing back to produce the 1980's most innovative car, the Voyager/Caravan minivans. In the 1990s, Chrysler has managed to reinvent itself, offering a range of cars that are truly contemporary, from the Neon, the Cirrus, and the big LH sedans to the incredible Viper.

Performance has always been associated with Chrysler, though this has not always been in keeping with Chrysler's conservative image. It seems as though there has always been a lunatic fringe within the company that somehow managed to get some pretty interesting automobiles and engineering features past management. Many examples exist, but the most well-known are the Hemi engines and the wild Daytona and SuperBird.

I have tried to include all the major high-performance collectibles produced since 1945. The Chrysler 300 letter series combined luxury and performance and set the standard in the fifties. As unique, but mostly unknown, are the DeSoto Adventurers. The Plymouth Fury was at one time (1956–58) the division's premier offering.

When it comes to the muscle car era, the Hemi is still king. Hemi-powered Dodges and Plymouths are still hot; no matter what the market does they remain the ultimate expression of sixties performance.

Obviously, originality is very important in the case of any collectible car. If you are new to the hobby, you'll find that there are many levels of originality, and much depends on how you are going to use the car. If you intend to show the car, be forewarned. A car is considered to be original only if it has been restored to factory specifications. It has to be in exactly the same condition as when it left the factory, using the same types and kinds of parts with which it was originally equipped. What about dealer installed equipment? This is a constantly debated, vast gray area, as is owner installed equipment. For example, fifties and sixties tire technology left a lot to be desired, and, logically, one would be tempted to fit the latest and safest rubber.

You'll find that the vast majority of enthusiasts have cars that are modified and are not "true" originals. However, the more original the car, the greater its worth. My personal view is that a car can be modified without reducing its inherent value if it can be returned to its original condition. Thus, if you install headers or an electronic ignition system, keep the factory stock parts so that if you decide to sell the car you can easily return it to stock. Simple modifications can be used to update these cars and make them much more enjoyable. A great deal of the enjoyment that can be derived from owning a collectible Chrysler is from driving it. The muscle cars in particular may seem crude when compared to the computerized techno-wizardry available today, but they sure get your heart beating a little bit faster.

As far as restoring a Chrysler, you'll find that there are many sources for engine and drivetrain components. The real difficulty in restoring Chryslers is finding interior pieces, trim, upholstery, body parts, and the like. Un- like other collectibles that you can practically rebuild by just calling an 800 number and charging all the parts you need to your credit card, you'll have to do quite a bit of legwork if you are trying to restore most Chryslers. There is no large reproduction industry serving the Mopar enthusiast, because most of the desirable cars were built in very small numbers. In my research, I found that there are many rare cars out there rotting away simply because there aren't any parts available to re- store them.

For this reason, it is very important that you try to find a car that requires the least amount of work. It may be tempting to buy a basket case, but without question, you'll find that you'll underestimate the final cost of the restoration. Your best bet is to join some of the Mopar clubs listed in the appendix, look at the want ads in the club magazines, and subscribe to *Hemmings Motor News*. You'll find that the asking prices in *Hemmings* can be all over the place for similar cars. However, under the parts headings, you'll get a fairly good idea of what parts cost and their availability.

You'll note that I haven't included a price guide. By the time this book is published it would be outdated. You can refer to the *Old Cars Price Guide* which is published several times a year, or *Hemmings*, but don't treat the asking prices you see as gospel. By the same token, really rare cars will rarely be advertised, but I guarantee you that if that person belongs to a club, someone in the club will know about it. You'll also find that parts dealers who specialize in Mopars are very knowledgeable about who has what for sale.

My goal in this book has been to present the most desirable high-performance Mopars from 1945-on. I hope the information and photos which follow will enable you to recognize and appreciate the best that Chrysler has had to offer.

Peter C. Sessler
Milford, Pennsylvania, November 1994

Investment Rating

★★★★★ These are the most sought after and also the rarest Mopars. They are already expensive and have the best potential for continued appreciation.

★★★★ Still rare and sought after, cars in this category were built in greater numbers than those rating five stars, but still are a solid investment.

★★★ Built in still greater numbers, the Mopars are more readily available and thus less expensive to acquire. This category also includes Mopars built in limited quantities but which have not yet caught on.

★★ These Mopars haven't caught on because they were built in great numbers and show no uniqueness. Relatively inexpensive to restore so be careful here.

★ This category includes any radically modified car, any race car and any car with the wrong engine.

Chapter 1

1955–65 Chrysler 300 Letter Series

★★★★	1955 C300
★★★★½	1956 300-B
★★★★★	1957 300-C
★★★★★	1958 300-D
★★★★½	1959 300-E
★★★★★	1960 300-F
★★★★	1961 300-G
★★★★	1962 300-H
★★★½	1963 300-J
★★★	1964 300-K
★★★	1965 300-L
★★★½	1964–65 300-K and 300-L convertibles

Before we explore the fabulous Chrysler 300s, let's define exactly what the 300s were. The 300 Letter Series, as they are called, were first introduced as a 1955 model, the C300. Subsequently, the 300-B came out as a 1956 model, the 300-C in 1957, and so on, until, in 1965 the last Letter Series, the 300-L, appeared (there was no 300-I). However, there are non-Letter 300s as well, beginning with the 1962 model year. Chrysler applied the 300 designation to more sedate models to improve sales. This practice continued until the 1971 model year. The last time Chrysler used the 300 name was in 1979. No doubt Chrysler will use the 300 name again in the future.

Because of the varied use of the 300 moniker, a fair amount of controversy exists within Chrysler circles as to what constitutes a 300. Dyed-in-the-wool purists recognize only the 1955–61 models as true 300s because these lived up to the concept of an exclusive racer built for the street. The Milestone Car Society also recognized these as Milestone cars, that is, modern classics, which tends to reinforce this concept. Others feel that true 300s include all the Letter cars ending with the 300-L. Still others feel that all Chrysler automobiles bearing the 300 name should be considered as true 300s.

I don't back any particular point of view, but I will say that this sort of controversy is essentially unresolvable, and that it is not unique to the 300s. For example, Shelby Mustang purists feel that only the 1965–67 versions are true Shelby Mustangs because the later 1968–70 models were designed and built by Ford.

Certainly the early 300 Letter cars are far more valuable yet the other 300s can provide just as much satisfaction.

1955 C300

The C300 is a car that happened to be introduced at exactly the right time. There are several reasons why Chrysler developed the C300, and at the same time, the C300 served as a focus for several elements that ensured its success and the success of the rest of the Chrysler line-up in 1955.

1955 Chrysler C300	
Engine	
Type	V-8
Bore x stroke, in.	3.81 x 3.63
Displacement, cubic in.	331
Compression ratio	8.5:1
Horsepower	300@5200 rpm
Torque	345@3200
Chassis and drivetrain	
Transmission	2-spd automatic
Front suspension	Independent, coil springs
Rear suspension	Live axle, leaf springs
Axle ratio	3.54:1
Brakes, front/rear	Drum/drum
General	
Wheelbase, in.	126
Height, in.	60.1
Width, in.	79.1
Length, in.	218.6
Weight, lb.	4,005

Certainly, engineering and utilitarian value are important factors that govern the success or failure of a particular car or car line. Chrysler products of the early fifties stressed these factors, featuring interior size versus exterior bulk. Yet styling sold cars more than any other factor. Chrysler sales fluctuated: 1951 and 1953 were good years, but 1954 proved to be a disaster with sales dropping over 450,000 units compared to 1953. By 1954, Chrysler styling was clearly not in favor with the public, which preferred the sleeker offerings from GM and Ford.

Fortunately, the Virgil Exner-designed 1955s arrived just in time. Exner had been Chryslers' chief stylist since 1953. The cars were sleeker, cleaner, longer, and much more attractive. The corporation dubbed the new styling as the "Forward Look." The driving public showed its approval by buying more Chrysler products than ever before. The C300 incorporated the best features of the Forward Look, and, as a top-of-the-line prestige model, it fit in quite nicely with the rest of the model line-up.

At the same time, an image car was needed to compete with the limited-production Chevrolet Corvette and the new Ford Thunderbird. The C300, however, was a high-performance image car, and nothing made in 1955 could touch it in terms of power. Performance is what had made the C300 famous, and that fame is due to Chrysler's most famous engine—the Hemi.

The first Hemi V-8, called the FirePower, was introduced in 1951 in Chrysler Imperials

The C300 used the Imperial grille and the New Yorker bumper. This particular car has owner-installed rear side-view mirrors. *Larry Bledsoe*

You only have to look at the 1955 offerings from other makes to see how elegant and simple the C300 was. Center car has the standard wheel cover, basically an Imperial unit with a checkered insert. Wire wheels came in sets of five. Center car has non-standard fog lights, and all have owner-installed rear side-view mirrors. *Larry Bledsoe*

The C300 dash. Note the awkward position of the shift lever. *Larry Bledsoe*

and New Yorkers, even though it had been in development on and off since 1935. It differed from other conventional overhead valve (ohv) designs because the combustion chamber was hemispherical, which all boiled down to greater efficiency, and that meant greater output. All things being equal, the Hemi design tended to produce more horsepower at higher rpm. Low-end torque and acceleration were about the same as with a typical wedge design, but as rpm increased, the Hemi outran the wedge.

Nineteen fifty-eight, unfortunately, was the last year for the first-generation Chrysler Hemi, but the Hemi design was used again by various Chrysler cars from 1966 to 1971. The only other domestic manufacturer to offer a production Hemi engine was Ford, which used the Boss 429 engine in a limited-production 1969–70 Mustang.

Soon after the FirePower's introduction, the engine found a home in Briggs Cunningham's C-2 roadsters that competed in the famous LeMans race in 1951. One of these cars managed to finish eighteenth. Carl Kiekhaefer, the Mercury Outboard Motor manufacturer, entered two Hemi-powered Chryslers in the 1951 Carrera PanAmericana. One car finished sixteenth. This and other race appearances generated considerable interest in the FirePower Hemi. From 1951 to 1953, more and more racers used this engine—even an Indy version was developed, but because it outpowered the established Offenhauser engine, last minute rule changes made it uncompetitive and it did not race.

The FirePower, then, was the natural choice for the C300. Displacement was set at

The 331ci FirePower V-8. Triangular air cleaner and valve covers were painted gold. *Larry Bledsoe*

331ci as offered in other Chrysler cars but with the use of two four-barrel Carter WCFB carburetors and a solid lifter camshaft that was close to the race 331s. Power was 300hp—the first American production engine with such a high output—which earned the 300 designation for the car.

As for the car itself, for cost reasons, it was based on the New Yorker two-door hardtop with rear quarters adapted from the Windsor. To lend prestige, the Imperial grille was used (as was the dash). Unlike other Chrysler offerings of 1955, the C300 wore minimal body ornamentation, to the point that no back-up lights or outside rearview mirrors were offered. Coupled with only a single color paint scheme limited to black, white, or red, all this simplicity tended to accentuate and define the C300s sleekness, which resulted in a determined yet graceful look. The C300 really stood out against the more garish styling typical of the day.

The interior was finished in tan leather and vinyl. Options included power steering, four-way power bench seat, power windows, radio, heater, tinted glass, and at a rather expensive $617, a set of five Kelsey-Hayes wire wheels that were also used on the 1954 Imperial. Air conditioning was not available. All C300s used a beefed-up two-speed automatic transmission, the PowerFlite, with a higher stall-speed torque converter. Standard axle ratio was 3.54:1, with the option of other ratios ranging from 3.36:1 to 4.10:1. Interestingly, the standard power brakes were drums all around, and the Imperial's four-wheel discs were not used on the C300. Higher-rate coil springs in front and leaves at the rear along

The 1956 300-B was essentially the same car, but the rear is more attractive.

The 1957 300-C was totally restyled, looking unlike any other American car. The mean-looking grille had a definite European flavor. The outside rear-view mirror was optional in 1957.

with heavy-duty shocks enabled the C300 to corner better than any other Chrysler offering in 1955.

For its time, the C300 was a fast car. The 0–60mph acceleration came in at about 10sec with a quarter-mile time in the high 17sec range, depending on axle ratio. A top speed of 130mph could be reached.

Appearing in January 1956 as a midyear introduction, which also coincided with the Daytona Speed Trials, the C300 was a fantastic success. It set many records and also won the Daytona Grand National race. By the end of the season, the C300 had won both NASCAR and AAA stock car championships, truly a formidable achievement for a new car. Credit for the stock car wins goes to Carl Kiekhaefer and his Mercury Outboard team.

Prospects

A total of 1,725 1955 C300s were built, and best estimates indicate that less than 200 survive today. Most that still exist are probably restored. Parts availability is poor, and if you do get a nonrestored or partially restored example, make sure that you have a source for parts. It is important not to forget what the C300 was in 1956: an exclusive automobile that most of the buying public could not afford. It was a car that had luxury, perfor-

Simple rearend treatment characterized the 300-C. One easy way to tell the difference between a 300-C and a 300-D from the rear is to look at the taillights. The 300-C's were much larger.

mance, engineering, and status, and it appealed to the same people who today buy Mercedes, BMWs, and Jaguars.

1956 300-B

Following the success of the C300, the next in the series was the 300-B, also introduced as a midyear model in January 1957. At a quick glance, the car seemed similar to the C300, yet it featured mechanical and styling differences.

Visually, the biggest difference was the restyled rear end, which had restrained (compared to later 300s') fins. These enhanced the sleekness and the appearance of the car. The front of the car, except for some minor detail differences, still used the Imperial grille. The rear taillights now incorporated back-up lights.

Mechanically, by increasing the bore by 0.13in, cubic displacement increased to 354 on the FirePower Hemi. Compression ratio was

For the first time, in 1957, the 300 was available as a convertible. Note upswept rear fins and large taillight lenses. The 300s also got red, white, and blue medallions on the rear fenders, interior, and grille. *Chrysler Corporation*

A few 1957 300-Cs were built with this headlight configuration. Quad headlights were not yet legal in all states, a situation which would change by 1958. *J.R. Beck*

1956 Chrysler 300B	
Engine	
Type	V-8
Bore x stroke, in.	3.94 x 3.63
Displacement, cubic in.	354
Compression ratio	9.0:1 (10.0:1 opt.)
Horsepower	340@5200 (355@5200 opt.)
Torque	385@3400 (405@3400 opt.)
Chassis and drivetrain	
Transmission	2-spd automatic (early), 3-spd automatic (late)
Optional transmission	3-spd manual
Front suspension	Independent, coil springs
Rear suspension	Live axle, leaf springs
Axle ratio	3.54:1
Brakes, front/rear	Drum/drum
General	
Wheelbase, in.	126
Height, in.	59.4
Width, in.	79.1
Length, in.	222.7
Weight, lb.	4,145
Performance	
0-60 mph	8.2
¼-mile e.t.@mph	NA
Source	*Motorsport*, 6/1956

increased to 9:1 from 8.5:1, and this resulted in 340hp. Later in the year, a 355hp version was made available. The extra horsepower was a result of the compression ratio having been raised to 10:1 and the use of a special 3in diameter exhaust system. The three-inch system was optional on the 340hp engine.

Just as important as the higher horsepower was the 300-B's ability to use the power for increased performance, due to the optional three-speed manual transmission (not available with power steering and brakes). Off-the-line acceleration was greatly improved: 0–60mph times in the low 8sec range were possible. Additional flexibility was added when the two-speed PowerFlite transmission was replaced by a cast-iron-case three-speed TorqueFlite automatic. Interestingly, Chrysler equipped only thirty-one cars with the three-speed manual. Other improvements included a 12-volt electrical system and the availability of air conditioning.

On the track, the 300s continued to perform well with the Kiekhaefer cars doing most of the winning. This was the last season that his cars would compete because he pulled out of circle-track racing the following year. In a sense, Kiekhaefer was too successful. There were times when his cars were booed because they usually won. His competitors constantly protested his cars yet never proved that he cheated. Rather, Kiekhaefer was a perfectionist, and teamwork and extremely careful preparation enabled his cars to beat everyone else's, including those of factory teams. An extremely proud and independent man, by the end of 1957, Kiekhaefer had had enough and withdrew from racing.

Prospects

Total production of the 300-B was 1,102. As with the C300, few remain. I have given the 300-B four-and-a-half stars—half a star more than the C300, because it is rarer and because its mechanical improvements and styling make it a better car. Current prices are about the same as for the C300, and you can definitely expect appreciation.

1957 300-C

The year 1957 brought the third 300 in the continuing series, the 300-C. Introduced to the public on December 8, 1956, it was very different from its predecessors—not only was the styling changed, but the car, along with the rest of the Chrysler line, featured significant engineering improvements as well.

The Flite Sweep show cars of 1955 influenced the styling, which meant fins—big, big fins. They were quite graceful and balanced, and they even looked functional. According to Chrysler, they added stability at speeds greater than 60–70mph. This was Virgil Exner at his best, and the industry looked to Chrysler for styling leadership, at least for the next two or three years.

The 392ci FirePower put out 375hp in standard form. As before, air cleaners and valve covers were painted gold. *J.R. Beck*

1957 Chrysler 300C	
Engine	
Type	V-8
Bore x stroke, in.	4.00 x 3.90
Displacement, cubic in.	392
Compression ratio	9.25:1 (10.0:1 opt.)
Horsepower	375@5200 (390@5400 opt.)
Torque	420@4000 (430@4200 opt.)
Chassis and drivetrain	
Transmission	3-spd automatic
Optional transmission	3-spd manual
Front suspension	Independent, torsion bars
Rear suspension	Live axle, leaf springs
Axle ratio	3.36:1
Brakes, front/rear	Drum/drum
General	
Wheelbase, in.	126
Height, in.	54.7 (55 convertible)
Width, in.	78.8
Length, in.	219.2
Weight, lb.	4,235 (4,390 convertible)
Performance	
0-60 mph	7.7
¼-mile e.t.@mph	NA
Source	*Motor Life*, 5/1957

There were two Derham-modified 300s built. One was a 300-B and the other was this 300-C. Modifications included the white canvas roof and the removal of all side trim. In the interior, Stewart-Warner instruments took the place of the originals. Derham, no longer in business, modified cars for the rich and famous. *Merle Wolfer*

The 300-C was still based on the New Yorker and its 126in wheelbase. The front of the car took on a decidedly mean look with its large trapezoidal grille. At that time, writer Bill Carroll coined the term "Beautiful Brute," and since then it has been used to describe all 300s.

In addition to the hardtop, a gorgeous convertible was available as well. Color choice expanded to five, with Parade Green and Copper Brown joining white, red, and black. The interior was still offered only in a tan leather. Special-order colors were available as well, and this was indicated by the numbers 888 on the car's data plate.

The FirePower Hemi was enlarged to 392ci and output increased to 375hp. The added displacement was achieved by boring and stroking: 4.00x3.90in. Increasing the stroke in any engine has the tendency to increase low-end torque at the expense of upper rpm capability, not a big factor on a street engine and certainly not on the FirePower. To help the Hemi breath even better, intake valve size was increased to 2in from 1.94in. The 300-C was still America's most powerful production car.

Transmission choice in 1957 was limited to the push-button-controlled three-speed TorqueFlite automatic standard with a three-speed manual optional. The three-speed manual was only available with the engine-chassis performance package.

The biggest engineering change that the 300-C shared with the rest of the Chrysler product line-up for 1957 was the introduc-

Above and right
The 1958 300-D featured restyled taillights. The grille was the same. The small, rectangular grille beneath the headlights provided fresh air to the front brakes for improved cooling. *Ken Driedger*

tion of torsion bar front suspension called Torsion-Aire. Coupled with other improvements such as a lower center of gravity, a wide lateral spring base, and a higher roll center, this made the 1957 Chryslers the best handling cars in the industry. The 300-C received torsion bars that were forty percent stiffer, providing race-car type ride and performance.

The major options included air conditioning, power steering, six-way power seat, power windows, radio, heater, tinted glass, and outside rearview mirrors. Various rear-axle ratios enabled the owner to fine-tune acceleration versus cruising capabilities. An optional exhaust system was available for the hardtop and utilized larger 2-1/2in pipes. The convertible was limited to a 2in system due to X-member interference.

The engine-chassis performance package included the 390hp version of the FirePower. The additional 15hp resulted from an increased compression ratio of 10.0:1 versus 9.25:1 and a hotter camshaft. The three-speed manual was available only with this package, and the lack of power steering and power brakes made the 300-C so-equipped a bear to drive. Only eighteen were made.

As far as performance goes, the 300-C accelerated quicker than its predecessor; yet at the Flying Mile at Daytona, it was slightly slower at 124.128mph. Top speed depended on rear-axle ratio—150mph was possible. Handling was considered excellent, but it is definitely dated by today's standards. The 9.00x14in tires were awful, and the drum brakes faded away under hard use. Still, few cars can rival its cruising capabilities and its seemingly effortless acceleration at higher speed.

Prospects

Production in 1957 increased to 1,767 hardtops and 484 convertibles. Of the two, the

Jerry Rushing's moonshine-running *Traveller* looked very stock, but in reality, was a highly modified car. The suspension was designed to sit at normal ride height with a full load of moonshine.

convertible is more desirable. As with the C300 and 300-B, you can expect appreciation in the years to come, especially with the convertible.

1958 300-D

The year 1957 proved so successful for Chrysler that few changes were made for the 1958 model year, and that included the 300-D. Some minor styling changes, some engine refinements, but basically it was the same car.

The D did not have the windshield visor that the C had, rear taillights were smaller, and the interior had differently styled upholstery and door panels. Standard color selection grew to six with additional special-order colors available. It seems to me, however, that white must have been the most popular color because most of the cars surviving today are white. The option list grew slightly, but the basic power options continued. As with the 300-C, the D's three-speed manual was available without power steering, power brakes, or air conditioning. Low production does make a manual-equipped 300-D rare, and that does add to its desirability. It is not a fun car to drive, though.

The 392ci FirePower's compression ratio was raised to 10.0:1 for a 380hp output. Cam timing was slightly subdued, which resulted

If you are going to run 'shine, you might as well do it in style. Switches to kill the taillights and operate an oil dump tank can be seen just below and left of the radio. Note the hefty seat belts.

in a smoother idle. The 390hp carbureted version was not available.

Instead, Chrysler tried to market a Bendix EFI (electronic fuel injection) version rated at 390hp. EFI is rather ho-hum today, yet in 1958, it was advanced—in fact, too advanced, as the system proved to be not only unreliable but also too expensive. Only sixteen cars were so equipped, and most of these were later converted to the two four-barrel carburetor setup. Chrysler has always prided itself as being, more than anything else, a company dedicated to engineering, and the EFI system was its answer to the mechanical injection system used on the 1957 Corvette, which was far more reliable.

Prospects

The 300-D has one distinction that makes it not only a desirable collector car but also unique. It is the last Letter Series that came with the Hemi engine. For cost reasons, later Letter Series cars were equipped with wedgehead engines. Some of these were faster, yet none have that indescribable something, the mystique of a Hemi.

The year 1958 was a bad one for the industry. Production of the 300-D was 618 hardtops and only 191 convertibles. For this reason, 300-D convertibles are currently somewhat more expensive than 300-C convertibles.

The 1958 300-D has a bit of folklore associated with it. Because Chryslers were so powerful and well built, they proved to be popular with moonshiners in the South. The most famous of these was the 300-D used by Jerry Rushing. With a full load of 'shine (adding at least 2,000lb to the car's weight), *Traveller*, as Rushing named his 300-D, could and did hit 140mph. Naturally the big

A highly modified 392 allowed *Traveller* to cruise at 140 mph.

1958 Chrysler 300D	
Engine	
Type	V-8
Bore x stroke, in.	4.00 x 3.90
Displacement, cubic in.	392
Compression ratio	10.0:1 (10.0:1 opt.)
Horsepower	380@5200 (390@5200 opt.)
Torque	435@3600 (435@3600 opt.)
Chassis and drivetrain	
Transmission	3-spd automatic
Optional transmission	3-spd manual
Front suspension	Independent, torsion bars
Rear suspension	Live axle, leaf springs
Axle ratio	3.31:1
Brakes, front/rear	Drum/drum
General	
Wheelbase, in.	126
Height, in.	55.2 (55.6 convertible)
Width, in.	79.6
Length, in.	220.2
Weight, lb.	4,305 (4,475 convertible)
Performance	
0-60 mph	8.4
¼-mile e.t.@mph	16.0@85
Source	*Road & Track*, 4/1958

Chrysler was modified with a reworked engine and suspension, and some of its "owner-installed" modification included a 20gal oil dump tank and a switch to turn off *Traveller's* brake lights. If a police car got too close, a flick of another switch (straight from James Bond) would spray a fine mist of oil on the road with predictable results. Rushing and *Traveller* were never caught.

Years later, Hollywood adapted Rushing's escapades for a television show, called "The Dukes of Hazzard." Rather than use rare 300-Ds, a Dodge Charger named the *General Lee* took the place of *Traveller*. *Traveller*, by the way, was the name of General Robert E. Lee's horse.

1959 300-E

The third year of the redesigned Chrysler line was 1959. Sales improved overall as the economy pulled itself out of the recession, yet sales did not grow appreciably at Chrysler. The finned revolution that Chrysler started was in full swing, with GM and Ford both bringing out some of the zaniest cars. Newness sold cars, and Chrysler would not have anything new until 1960.

The 300-E featured a slightly different grille pattern and different hubcaps, but the most obvious change was the rear-end treatment. Overall, the design impressed people as being busier, but that was because the 300-E shared the same body as the New Yorker. However, the E was by far the cleanest looking Chrysler. No heavy ornamentation here.

New features included such things as an optional Mirromatic rearview mirror that adjusted itself according to the rear backlight and breathable "Living Leather." An interesting feature was the swivel front seats that were supposed to facilitate entry and exit. The options list grew with items such as the automatic headlamp dimmer, Auto-Pilot (cruise control carried over from 1958), and True-Level Torsion-Aire air suspension. However, most optional axle ratios were eliminated—only a 2.93:1 ratio was available while standard was a 3.31:1. The optional 2-1/2in exhaust system, manual transmission, and manual steering and brakes were also eliminated.

Power steering, power brakes, and the three-speed TorqueFlite automatic transmission were all standard equipment. Color choices remained at six (black, white, red, gray, tan, and copper), and the interior was again finished in tan. Special-order colors and interiors were available as well.

1959 Chrysler 300E	
Engine	
Type	V-8
Bore x stroke, in.	4.18 x 3.75
Displacement, cubic in.	413
Compression ratio	10.1:1
Horsepower	380@5000
Torque	450@3600
Chassis and drivetrain	
Transmission	3-spd automatic
Front suspension	Independent with torsion bars
Rear suspension	Live axle, leaf springs
Axle ratio	3.31:1
Brakes, front/rear	Drum/drum
General	
Wheelbase, in.	126
Height, in.	55.3 (55.7 convertible)
Width, in.	79.5
Length, in.	220.2
Weight, lb.	4,290 (4,350 convertible)
Performance	
0-60 mph	8.7
¼-mile e.t.@mph	17.2@92
Source	*Sports Car Illustrated*, 8/1959

The 1959 300-E featured a restyled grilled insert and a restyled rear-end treatment. *Guy Morice*

The biggest change on the 300-E was the replacement of the 392 FirePower Hemi with a 380hp 413ci Golden Lion wedge-head V-8. It featured dual Carter AFB carburetors and a camshaft that was a bit milder than the Hemi's. Low-end acceleration was better than that of the previous year's Hemi-powered 300-D, but like most wedge engines, the 300-E ran out of breath at higher rpm. At Daytona, the best that an E could do was a lackluster 120.481mph.

Still, the 300-E was a better street car than any of its predecessors—quieter, smoother, and far more tractable. Unfortunately, it failed to enthuse the buying public as production reached a low point of 550 hardtops and 140 convertibles. Of course, a list price of more than $6,000 for an optioned-out 300-E hardtop and of close to $6,500 for the convertible limited sales because the cars just weren't that different from other Chrysler offerings. It must also be remembered that the new four-seater Thunderbird was selling well and that it was less expensive.

Prospects

Chrysler lost its styling leadership after 1959. The market moved away from fins, and the styling that Chrysler products had in the early sixties was, with a few exceptions, awful.

That is why we can expect the 1955–59 300s to appreciate. They represent a period when the Chrysler Corporation was definitely the styling and engineering leader in the industry.

Be aware that the 1957–59 Chryslers had the reputation of being rustbuckets. Be extra careful when you are considering investing in one of these beauties. Parts availability is poor,

The big 413 wedge replaced the Hemi. *Guy Morice*

Above and right
The 1960 300-F is considered among the most desirable of all 300s. Continental spare, an Exner touch available on other Chrysler cars for several years, came with the 300-F.

particularly for parts unique to the 300s. You can still find unrestored examples, and the reason they are still unrestored is lack of parts.

1960 300-F

January 15, 1960, saw the introduction of the 300-F, the sixth in the series. The horsepower race was on again: the 300-F could be had with a 400hp engine and a four-speed manual transmission. Chrysler got a lot of advertising mileage out of that offering because, according to most sources, it only built seven cars with the 400hp engine. But they sure performed.

For example, the 300-F bettered the Flying Mile record at Daytona (which previously stood at 139.37mph set by a 300-B) with a two-way average of 144.927mph and a one-way run at 147.783mph. For some serious numbers, Andy Granatelli entered a supercharged 300-F at Bonneville and set a record of 184.049mph and a one-way run at 189.990mph. If you consider the poor aerodynamics of the 300-F and its weight, these are tremendous figures.

The year 1960 was one of many changes. Most important was the introduction of unit construction or unibody to the entire Chrysler line, except for the Imperial. Simply put, rather than being the traditional body-on-

frame construction, the body and chassis were now one welded unit to which the suspension and engine were attached. Advantages included a 100 percent improvement in rigidity which resulted in a tighter, quieter ride and ease of manufacture. However, rust was a much more serious problem in a unibody design, and this rusting out weakened the car's structural integrity.

Other engineering changes involved the 413ci Golden Lion V-8. Although, as in 1959, two Carter AFB four-barrels were used, these were mounted on a unique ram induction setup designed to boost midrange torque. By shortening or lengthening tube length, Chrysler engineers were able to tailor to the engine's performance characteristics. The goal was to provide more power in the midrange for better passing. A thirty-inch-long tube was found to be optimum for these parameters. The standard engine thus produced 375hp at 5200rpm with 495lb-ft of torque also at

1960 Chrysler 300F	
Engine	
Type	V-8
Bore x stroke, in.	4.18 x 3.75
Displacement, cubic in.	413
Compression ratio	10.1:1 (10.1:1 opt.)
Horsepower	375@5000 (400@5200 opt.)
Torque	495@2800 (465@3600 opt.)
Chassis and drivetrain	
Transmission	3-spd automatic
Optional transmission	4-spd manual
Front suspension	Independent, torsion bars
Rear suspension	Live axle, leaf springs
Axle ratio	3.31:1
Brakes, front/rear	Drum/drum
General	
Wheelbase, in.	126
Height, in.	55.3 (55.7 convertible)
Width, in.	79.5
Length, in.	219.6
Weight, lb.	4,270 (4,310 convertible)
Performance (with 400 hp engine)	
0-60 mph	7.2
¼-mile e.t.@mph	NA
Source	*Motor Life*, 6/1960

2800rpm.

The long ram tubes did have a drawback: They severely limited upper rpm performance. By having cut the internal length of the tube to 15in, the torque peak was raised to 3600rpm on the optional 400hp version. This enabled the engine to keep on pulling to 5200rpm. Other improvements included larger exhaust valves (1.74in versus 1.60in), a solid lifter camshaft, and a freer-flowing 2-1/2in exhaust system. Acceleration times for 0–60mph came in the low 8sec range and, for the quarter-mile, in the low 16sec range for the 375hp equipped 300-F. While the 400hp version took about a second off these times.

The standard transmission was the three-speed TorqueFlite with a 3.31:1 final drive ratio. The optional four-speed was only available as a package with the 400hp engine. This was the Pont-a-Mousson aluminum unit used in the French Facel Vega, which had a Chrysler Hemi engine. An expensive $800 option on an already expensive car (about $6,000 with other options) limited its appeal and availability to seven units, including one convertible. But as I said before, it sure made good copy!

The styling, although showy, was dominated by the large, flared rear fins, a Continental-type trunk lid (dubbed the toilet seat by enthusiasts) and a redesigned, serious-looking grille. Color choices were reduced to four: Formal Black, Toreador Red, Alaskan White, and Terra Cotta. Other exterior and interior colors were available on a special-order basis.

Above and right
The 300-F interior used four bucket seats with tan leather standard. The redesigned dash was a bit on the flashy side.

Prospects

In spite of great styling and performance, the 300-, was not a great seller. Chrysler produced 964 hardtops and 248 convertibles. Few remain with the convertible leading the way in terms of appreciation.

You can expect 300-F prices to rise because the 300-F embodied the best of Virgil Exner's styling with great performance. Indeed, the 300-F is a large, dominating car. Even with the ignition off, the 300-F exudes a sense of raw power which is confirmed when the engine is turned on.

1961 300-G

In 1961, the 300-G appeared, a Letter car that was not too different from the 300-F. The most noticeable change was the front grille; basically the 300-F grille was turned upside down and the quad headlights were slanted inward as on the 1960 Lincoln. The grille change diluted some of the tough-guy look,

1961 Chrysler 300G	
Engine	
Type	V-8
Bore x stroke, in.	4.18 x 3.75
Displacement, cubic in.	413
Compression ratio	10.1:1
Horsepower	375@5000
Torque	495@2800
Chassis and drivetrain	
Transmission	3-spd automatic
Optional transmission	3-spd manual
Front suspension	Independent, torsion bars
Rear suspension	Live axle, leaf springs
Axle ratio	3.31:1
Brakes, front/rear	Drum/drum
General	
Wheelbase, in.	126
Height, in.	55.6 (56 convertible)
Width, in.	79.4
Length, in.	219.8
Weight, lb.	4,260 (4,315 convertible)
Performance (with 3-spd manual)	
0-60 mph	8.3
¼-mile e.t.@mph	NA
Source	*Motor Trend*, 6/1961

although the Continental spare had disappeared.

Still, the 300-G was the last example of the waning Exner Forward Look fins. By this time, Chrysler had lost the styling leadership earned in the late fifties, and some of the 1961 models were truly "different," to say the least.

Basically, the 300-G was a carryover. The 375hp Long Ram engine was standard, but the 400hp Short Ram now came with a three-speed manual (non-synchro first) that was also available on the Lancers and Valiants. We do not know how many 300-Gs so equipped were built, but it cannot have been more than a handful.

One interesting feature on all Chrysler V-8s was the switch from a generator to an al-

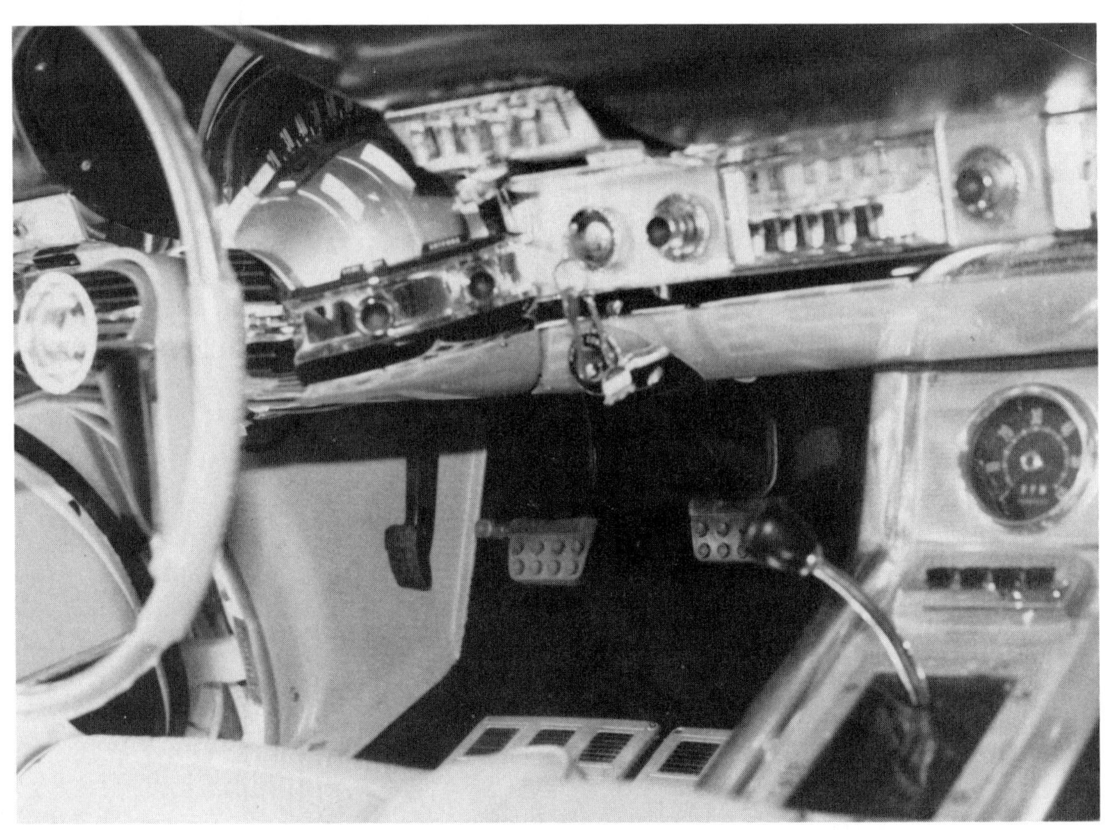

Rare four-speed convertible 300-F. Rev counter's low position in the center console made it almost useless. *Tom Turner*

ternator; in 1960, only the 400hp engine had been equipped with an alternator. Also wheel size was increased from 14in to 15in, which improved handling.

The interior was still finished in tan leather, although the pattern was somewhat different from that used in the 300-F. Exterior colors were limited to four: Formal Black, Mardi Gras Red, Cinnamon, and Alaskan White.

Prospects

Production reached 1,280 for the hardtop and 337 for the convertible. These low numbers ensure the G's collectibility, but it is also important to remember that the 300-G was the last Letter car that did not share the 300 designation with other Chryslers.

1962 300-H

Many changes were made for 1962; most obvious were those in the styling department with an end to the fins. But the front grille design was basically a carryover from the 300-G, as was the interior. The 300-H was now built on a shorter 122in wheelbase from the Newport rather than on the 126in New Yorker platform. This resulted in improved performance as the H was about 300lb lighter, although it can be argued that the ride suffered somewhat.

Performance, however, was excellent. The standard engine was a 380hp 413 featuring inline-mounted dual four-barrel Carters, providing better than the usual Letter Series performance: 0–60mph in the high 7sec range and the quarter-mile time in the low 15s.

For the performance enthusiast, a rare 405hp version of the 413 was available as a

Long Ram 413 Wedge provided 375 healthy horses.

The 300-G featured grille and rearend styling changes. *Tom Turner*

1962 Chrysler 300H	
Engine	
Type	V-8
Bore x stroke, in.	4.18 x 3.75
Displacement, cubic in.	413
Compression ratio	10.0:1
Horsepower	380@5200
Torque	485@3200
Optional engine	405 hp 413 cid (dealer-installed)
Chassis and drivetrain	
Transmission	3-spd automatic
Front suspension	Independent, torsion bars
Rear suspension	Live axle, leaf springs
Axle ratio	3.23:1
Brakes, front/rear	Drum/drum
General	
Wheelbase, in.	122
Height, in.	55.2
Width, in.	79.4
Length, in.	214.9
Weight, lb.	4,240

dealer-installed option. This was a Short Ram version with a nasty solid lifter cam and an 11.1:1 compression ratio. Even on a warm day, it took about fifteen minutes of idling (or trying to keep it idling) before it ran on its own. Performance was impressive: 0-60mph in the high 6sec range and quarter-mile times in the 14.7 range.

A modified 300-H with the 405hp engine was entered in the 1962 Winternationals. Best times included an elapsed time of 12.88sec and a speed of 108.40mph. This particular car had the TorqueFlite transmission, a 4.56:1 axle, headers, and the usual drag-racing modifications to encourage proper weight transfer.

An Andy Granatelli 300-H turned 189.9mph at Bonneville with the help of two

McCulloch Supercharges and also set the Flying Mile record at 179.472mph. Very fast indeed.

Standard transmission was the TorqueFlite automatic with a three-speed manual optional. Either transmission was available with the 405hp engine as it was a dealer-installed option.

The interior was finished in the usual tan leather, but other colors were available under special order.

Prospects

Sales hit an all time low with the 300-H: 435 hardtops and only 123 convertibles. I doubt if more than seventy-five 300-Hs have survived. If you are interested in a 300-H with the 405hp engine, you must make sure that it is an original dealer-installed unit, and that means documentation. Look for appreciation to be quite strong simply because so few have survived, especially in the convertible.

1963 300-J

The 1963 Chrysler line was totally redesigned under the direction of Elwood Engel, who replaced Virgil Exner. Engel, hired from Ford, was responsible for the restyled 1961 Lincoln and Thunderbird. The new Chrysler look was dubbed the "Crisp, Clean, Custom Look." Still, the Exner influence was there—for example, the large trapezoidal grille, and large wheelwell cutouts—and it wasn't until the 1965 model year before Exner's influence finally vanished.

The interior, save for some minor trim changes, was essentially unchanged on the 300-G. *Tom Turner*

The 300-H, shorn of fins, looked very similar to the 300-G. This particular car has the rare, dealer-installed 405hp Short Ram 413 and Special Order (code 888) black leather interior.

1963 Chrysler 300J	
Engine	
Type	V-8
Bore x stroke, in.	4.18 x 3.75
Displacement, cubic in.	413
Compression ratio	9.6:1
Horsepower	390@4800
Torque	485@3600
Chassis and drivetrain	
Transmission	3-spd automatic
Front suspension	Independent, torsion bars
Rear suspension	Live axle, leaf springs
Axle ratio	3.23:1
Brakes, front/rear	Drum/drum
General	
Wheelbase, in.	122
Height, in.	55.6
Width, in.	79
Length, in.	215.5
Weight, lb.	4,235
Performance (with 400 hp engine)	
0-60 mph	7.5
¼-mile e.t.@mph	15.4@94
Source	*Car and Driver*, 4/1963

The 300-H interior.

The 300-H convertible is very rare car with only 123 built. *Ron Wenger*

The 300-J continued the tradition that started with the C300: a powerful engine mated to a roadworthy chassis luxuriously appointed. The 300-J still came with a leather interior as standard equipment and with five exterior color choices: Formal Black, Alabaster, Madison Grey, Claret, and Oyster White.

Performance was good too, with a 390hp 413 featuring the Short Ram Induction setup. No optional engine was available, but a three-speed manual was an option in lieu of the standard heavy-duty Torqueflite automatic.

However, to the dismay of the motoring press, the suspension was softened as can be seen by the table below.

Ride rate at wheel	**300-G**	**300-H**	**300-J**
Front, lb-in	160	130	125
Rear, lb-in	190	160	150

Soft suspension tends to produce a more comfortable ride at the expense of cornering capability. However, you can easily compensate for softer springs by installing much heavier shock absorbers. In fact, current guidelines for combining handling and ride call for relatively soft springs with firm shocks and large diameter front and rear sway bars. The cars listed in the chart all came with a 0.75in front sway bar.

Firmer shock absorbers should not hurt originality for those enthusiasts who still enjoy driving their cars, and the same can be said of modern tires and rear sway bars, unless the car is strictly a show car.

There were some changes made to the J's brakes; Bendix brakes replaced Chrysler's center-plane brakes and proved to be more resistant to fade. A foot pedal actuated the parking brakes.

Prospects

Although the tradition continued with the 300-J, sales hit an abysmal low of 400 units, all hardtops. Perhaps the $1,700 differential between the J and a non-letter 300 adversely affected sales, and Chrysler's lack of promotion certainly did not help.

In spite of the fact that so few 300-Js remain, the prospect for appreciation is modest simply because the Letter Series by 1963 had lost its exclusivity and prestige. If the earlier Letter cars rise in value, so too will the 300-J, but don't expect the J to rise simply on its own.

1964 300-K

The 1964 300-K made a comeback, at least in terms of sales: 3,022 hardtops and 625 convertibles. This success probably reflected a $1,000 reduction of the list price. The leather interior was replaced by vinyl (leather was optional) and the standard 413 engine featured a single four-barrel. Chrysler also advertised the car more extensively.

The standard 413 pumped out 360hp while the optional Short Ram-equipped K put out 390 horses. Wheel size reverted to a 14in rim with an 8.00x14in standard tire. On the Short Ram engine, tires were 8.50x14s, which

Short Ram 405hp 413. Tall air cleaner studs were designed to accommodate the taller air cleaner elements required with the 405. Power brake booster was relocated in the cavity of left front fender.

1964 Chrysler 300K	
Engine	
Type	V-8
Bore x stroke, in.	4.18 x 3.75
Displacement, cubic in.	413
Compression ratio	9.6:1
Horsepower	360@4600
Torque	470@3200
Chassis and drivetrain	
Transmission	3-spd automatic
Optional transmission	4-spd manual
Front suspension	Independent, torsion bars
Rear suspension	Live axle, leaf springs
Axle ratio	3.23:1
Brakes, front/rear	Drum/drum
General	
Wheelbase, in.	122
Height, in.	55.3
Width, in.	79
Length, in.	215.3
Weight, lb.	4,250

were an option on the lower horsepower engine as well.

Interestingly, Chrysler made available a new four-speed manual transmission; however only eighty-four cars were equipped with it. The shift lever was uniquely located: it was on the floor to the left of the console rather than at the center of the console. The standard TorqueFlite no longer used push buttons; instead, the driver selected gears with a console-mounted shifter. Both the automatic and manual transmissions used the same final drive ratio, 3.23:1.

A 120mph speedometer replaced the 150mph unit, and the tachometer became optional. A tilt steering wheel was offered for the first time, and color choices expanded to sixteen.

Above and right
A 1963 300-J with horizontal headlights. Still sleek and elegant without any side adornments. *Don Drakulich and Ron Wenger*

Prospects

From the enthusiast's point of view, all these changes represented a further watering down of the 300's original concept. Still, it was a better car than the non-Letter 300, and in spite of the fact that more 300-Ks were built than any other Letter Series, few remain.

A 300-K convertible with optional leather and 390hp engine is probably the most desirable from a collector's point of view. Look for clean, restored examples rather than investing in a K that needs restoration. Parts availability is poor. Although the 300-K has less glamour associated with it than do earlier 300s, it is a solid investment because, from mechanical and styling points of view, it still has the luxury and performance associated with a Letter Series car, and just as important, a K bears the Virgil Exner imprint.

1965 300-L

There were many reasons for the demise of the Letter Series. Sales were never that good, and Chrysler could not have made much money on the line. Since 1962 when the non-Letter 300s appeared, the impact of the Letter Series as a means of increasing sales had diminished considerably. Performance did sell cars, but by 1962 the emphasis was on the intermediates. It seemed that if you were a car manufacturer, all you had to do was put your biggest, most powerful engine in your least expensive platform, give this creation a catchy name and watch sales take off. Luxury options made the Letter cars too expensive, and thus out of range of the exploding youth market. The Letter Series had the reputation of being big highway cruisers. Now, quarter-mile times were more important.

A 1964 300-K convertible.

The 300-K interior was dominated by the large, squarish steering wheel.

1965 Chrysler 300L	
Engine	
Type	V-8
Bore x stroke, in.	4.18 x 3.75
Displacement, cubic in.	413
Compression ratio	10.1:1
Horsepower	360@4800
Torque	470@3200
Chassis and drivetrain	
Transmission	3-spd automatic
Optional transmission	4-spd manual
Front suspension	Independent, torsion bars
Rear suspension	Live axle, leaf springs
Axle ratio	3.23:1
Brakes, front/rear	Drum/drum
General	
Wheelbase, in.	124
Height, in.	55.3
Width, in.	79.5
Length, in.	218.2
Weight, lb.	4,660
Performance	
0-60 mph	8.8
¼-mile e.t.@mph	17.3@82
Source	*Motor Trend,* 3/1965

Only eighty-four 300-Ks came with a four-speed manual transmission. *Tom Smith*

The 1965 300-L was based on a completely redesigned body. It was handsome but bore little resemblance to previous Letter cars. Engine choice was limited to a single four-barrel 413 rated at 360hp with either a standard TorqueFlite automatic or an optional four-speed manual. In either case, the 3.23:1 rear was the only one available. Vinyl was the standard material for the interior, while leather was optional. Car buyers chose from many exterior colors, and for those so inclined, a full vinyl roof was available.

Gone were the days of Daytona and the Speed Trials. The 300-L could go from 0–60 mph in the high 8sec range, respectable for a full-size car, but nothing more.

Prospects

Production was relatively high: 2,405 hardtops and 440 convertibles. You can expect these 300-Ls to appreciate, although perhaps not quite as rapidly as do earlier Letter Series, but they are definitely the most desirable 1965 Chryslers.

1956–58 Plymouth Fury

If you are looking for a fifties Plymouth with lots of pizzazz, there is really only one that you can consider. The 1956–58 Furys were distinctive automobiles that combined excellent performance with a unique and exciting visual package. The Fury was to Plymouth what the 300 was to Chrysler.

Above and right
The 1956 Plymouth Fury added some excitement to the usual staid Plymouth lineup. *Thomas Ray*

1956–58

Capitalizing on the momentum created by the revitalized styling of 1955, Plymouth brought out the Fury as a midyear introduction on January 10, 1956. Based on the two-door hardtop Belvedere, it was nevertheless distinctive. The one available color, an off-white, was set off effectively by a gold anodized bodyside sweepspear. Gold was also used in the grille and on the wheel covers. All 1956–58 Furys sported this color combination.

Unlike its Chrysler 300-B and DeSoto Adventurer counterparts, the Fury was not powered by a Hemi-head engine. Instead, a 303ci Hy-Fire V-8 with polyspheric combustion chambers was used. In this kind of combustion chamber design, the cylinder head is relatively shallow and the piston design de-

1956 Plymouth Fury	
Engine	
Type	V-8
Bore x stroke, in.	3.81 x 3.31
Displacement, cubic in.	303
Compression ratio	9.25:1
Horsepower	240@4800
Torque	310@2800
Chassis and drivetrain	
Transmission	3-spd manual
Optional transmission	2-spd automatic
Front suspension	Independent, coil springs
Rear suspension	Live axle, leaf springs
Axle ratio	3.73:1
Brakes, front/rear	Drum/drum
General	
Wheelbase, in.	115
Height, in.	58.8
Width, in.	74.6
Length, in.	204.8
Weight, lb.	3,650

termines the combustion chamber shape. A rating of 240hp was reached through the use of a 9.25:1 compression ratio, a solid lifter cam, four-barrel carburetor, and a free-flowing exhaust system. Naturally, the Fury came with heavy-duty springs and shocks, but the Fury sat 1in lower than other Plymouths. An inch may not be much, but it makes a big difference in the way a car looks, lending a purposeful appearance.

A preproduction Fury ran at Daytona that year with a Flying Mile time of 124mph. On the street, the Fury could reach 60mph in the mid-9sec range and the quarter-mile time in the low 17s. A three-speed manual was standard, and the two-speed PowerFlite automatic was optional. Other optional equipment included power brakes, steering, seat and windows, and air conditioning.

1957 Plymouth Fury	
Engine	
Type	V-8
Bore x stroke, in.	3.91 x 3.31
Displacement, cubic in.	318
Compression ratio	9.25:1
Horsepower	290@5400
Torque	325@4000
Chassis and drivetrain	
Transmission	3-spd manual
Optional transmission	3-spd automatic
Front suspension	Independent, torsion bars
Rear suspension	Live axle, leaf springs
Axle ratio	3.73:1
Brakes, front/rear	Drum/drum
General	
Wheelbase, in.	118
Height, in.	53.5
Width, in.	79.4
Length, in.	206
Weight, lb.	3,595

The 240hp version of the 303ci Hy-Fire V-8 was standard equipment. *Thomas Ray*

The dash, by 1950 standards, was relatively simple. This Fury had optional air conditioning. *Thomas Ray*

A 290hp 318 with dual quads made the Fury the quickest 1957 Plymouth. *Dale Frahm*

For 1957, the Fury had a much cleaner, lighter grille. The rather elegant side spears contributed to the car's sleekness while emphasizing the rear fins. *Dale Frahm*

1958 Plymouth Fury	
Engine	
Type	V-8 (carbureted 350 V-8 opt.)
Bore x stroke, in.	3.91 x 3.31 (4.06 x 3.38 opt.)
Displacement, cubic in.	318 (350 opt.)
Compression ratio	9.25:1 (10.0:1 opt.)
Horsepower	290@5400 (305@5000 opt.)
Torque	325@4000 (370@3600 opt.)
Chassis and drivetrain	
Transmission	3-spd manual
Optional transmission	3-spd automatic
Front suspension	Independent, torsion bars
Rear suspension	Live axle, leaf springs
Brakes, front/rear	Drum/drum
General	
Wheelbase, in.	118
Height, in.	53.5
Width, in.	79.4
Length, in.	206
Weight, lb.	3,510

Perhaps the reason the Fury was not equipped with a Hemi was that it would have blown the doors off of the 300-B. Built on a 115in wheelbase, the Fury weighed hundreds of pounds less than the 300-B, and you just couldn't have a Fury that was faster than the car that was the company's most prestigious and one of its most expensive. Plymouth built and sold low-priced cars, and the Fury's base price was $2,866, which helps to account for sales of 4,485 units.

If 1956 was successful, then the 1957 was an absolute smash. Chrysler restyled the entire line under the eye of Virgil Exner and took the styling lead from General Motors. The 1957 Fury was longer, lower, wider, and much sleeker than the 1956.

For 1958, the Fury was relatively unchanged. Note the delicate, almost frail-looking roof. *Norman Buelow*

Besides the new styling, the Fury benefited from the torsion bar front suspension and optional TorqueFlite automatic transmission. Optional equipment remained nearly the same as in 1963, but the 303 was now bored to 318ci and named the V-800. Two four-barrel carbs helped to raise horsepower to 290, and performance improved accordingly. Sales too hit a high of 7,438.

The 1958 version was restyled slightly including four headlights, matching underbumper grille, and smaller taillights. The V-800 was standard while a new Golden Commando 350ci wedge was optional. Rated at 305hp, it too sported two four-barrel carburetors.

A 315hp version was also available; it used fuel injection, the same ill-fated system that was used on the 300-D. Most, if not all were later converted to carburetors. Production for the 1958 reached 5,303.

The year 1959 marked the end of the Fury as a limited-production high-performance automobile. The name Fury was used on an entire line, although a Sports Fury model was available as a hardtop and convertible, but that did not last into 1960.

Prospects

Prospects are good for the 1956–58 Furys. Relatively unknown, they have been over-shadowed by the 300 Letter Series, so

Front grille featured some minor changes—most obvious was the lower grille's horizontal bars. The taillights, too, were changed slightly. Both 1957 and 1958, Furys featured beautifully styled bumpers, unlike the battering rams of today's cars. *Norman Buelow*

Interior was functionally simple. This Fury has a manual transmission. *Norman Buelow*

you should be able to find a nice example. Although Furys lack the exclusivity of a 300, they are prime examples of Virgil Exner at his best, and for this reason you can expect strong appreciation. Despite the relatively high production run, not many Furys have survived into the 1990s, and as with all late fifties Chrysler products, watch out for rust.

The 1958 290hp 318ci came with gold-painted oval air cleaners mounted on the carburetors. Optional engine for the Fury was a 305hp 350ci B-block. *Norman Buelow*

Chapter 3

1956–60 DeSoto Adventurer

★★★★★	**1956 Pacesetter**
★★★★★	**1956–59 Adventurer convertible**
★★★★	**1956–59 Adventurer sedan**
★★★	**1960 Adventurer**

In the marketing scheme, the DeSoto was Chrysler's Oldsmobile, competing in the medium-price field. GM consisted of Chevrolet, Pontiac, Oldsmobile, Buick, and Cadillac. Chrysler's complementary offerings were Plymouth, Dodge, DeSoto, Chrysler, and Imperial. Sales in the fifties were decent; in some years DeSoto outsold Chrysler. However, un-

Above and right
The 1956 Pace Car convertible put the spotlight on DeSoto. Note prominent dual exhausts and triple taillights. *Gus DeGazio*

1956 DeSoto Adventurer	
Engine	
Type	V-8
Bore x stroke, in.	3.78 x 3.80
Displacement, cubic in.	341
Compression ratio	9.25:1
Horsepower	320@5200
Torque	NA
Chassis and drivetrain	
Transmission	3-spd automatic
Front suspension	Independent, coil springs
Rear suspension	Live axle, leaf springs
Axle ratio	3.73:1
Brakes, front/rear	Drum/drum
General	
Wheelbase, in.	126
Height, in.	58
Length, in.	220.9
Weight, lb.	3,870

Single 4-barrel FireFlite provided plenty of oomph. *Gus DeGazio*

The 1956 Adventurer hardtop. Gold anodized wheel covers set it off. The 1956 Plymouth Fury used the same wheel cover, but without the DeSoto script. *Larry Zappone*

like its sister divisions, DeSoto was not able to bounce back from the 1958 recession. The year 1957 saw 126,514 DeSotos produced; in 1958, production dropped to 49,445; and by 1960, a dismal 26,081 units were built. An abbreviated 1961 model run saw an additional 3,034 units. Without any great fanfare, an announcement on November 18, 1960, indicated that production would cease by month's end. The DeSoto had lost its appeal.

Still, from a collector's point of view, there are some bright moments, namely the Adventurer, introduced as a midyear entry on February 18, 1956, and the Pacesetter convertible that preceded the Adventurer by about a month.

1956 Pacesetter

DeSoto was selected to pace the 1956 Indianapolis 500 race. As now, this meant lots of publicity, and to take full advantage of it, DeSoto produced a special Pacesetter convertible. Based on the Fireflite convertible, the Pace cars were available only in white accented by lots of gold: gold wheel covers, gold top and side sweeps, gold vinyl upholstery, gold instrument panel, ivory and gold steering wheel, and gold threads in the black carpets. Even in the age of flash, this combination stood out.

The powertrain consisted of a DeSoto 225hp 330ci Hemi sporting a single four-barrel carburetor. The PowerFlite automatic transmission was standard equipment. The rest of the car, mechanically, was standard Fireflite.

Prospects

Most Pacesetter convertibles were equipped with luxury options such as power seats and power windows. Most sources list that between 100 and 200 examples were built. According to the National DeSoto Club,

The 320hp 341ci Hemi engine was good for over 135mph. *Larry Zappone*

The 1957 DeSoto Adventurer embodied all the elements of Exner's "Forward Look," while retaining the Adventurer's distinctive glittery look. *Alan Linsky*

1957 DeSoto Adventurer	
Engine	
Type	V-8
Bore x stroke, in.	3.80 x 3.80
Displacement, cubic in.	345
Compression ratio	9.25:1
Horsepower	345@5200
Torque	355@3600
Chassis and drivetrain	
Transmission	3-spd automatic
Front suspension	Independent, torsion bars
Rear suspension	Live axle, leaf springs
Axle ratio	3.73:1
Brakes, front/rear	Drum/drum
General	
Wheelbase, in.	126
Height, in.	55
Width, in.	78.2
Length, in.	218
Weight, lb.	4,040

however, the true figure may be closer to 400. Incredibly, few have survived; not more than twenty-five or so are known to remain. Definitely a good bet for appreciation.

1956–60 Adventurer

The Adventurer was similar to the Pace cars in terms of styling; however, it was only available as a hardtop in 1956. Color choice was more varied: three colors, gold, black, and white, were available in six combinations.

Considerably more powerful, the Adventurer used a dual four-barrel 341ci Hemi that pumped out 320hp, only 20hp less than the standard Chrysler 300-B engine. The Adventurer's engine benefited from a 9.25:1 compression ratio, a high-lift camshaft, larger cylinder head ports, and performance-oriented intake manifold and ignition timing. It was a strong, durable engine. A prototype ran 137mph at Daytona, and later the same car ran 144mph at the Chrysler Proving Grounds. The Adventurer provided 300-B performance, perhaps not in the same understated way, but at a much lower cost. Production in 1956 was a low 996 units.

DeSoto shared in all the improvements made for the 1957 model year: TorqueFlite

Above and right
The 1957 DeSoto Adventurer embodied all the elements of Exner's "Forward Look," while retaining the Adventurer's distinctive glittery look. *Alan Linsky*

WQE 623

The 1958 Adventurer featured revised sidespear paint which gave the car a lighter, sleeker look. This is one of the few convertibles still in existence. *Jim DeGregorio*

For 1959, the Adventurer was available only in white or black and featured a restyled grille. *Galen Erb*

transmission, front torsion bar suspension, and, of course, styling. DeSoto benefited from the masterly touch of Virgil Exner, who in that year was elevated to the newly created position of vice president, director of styling.

The Adventurer series looked great, especially as a convertible (in addition to the two-door hardtop). Also befitting the flagship that it was, the Adventurer received a 345ci Hemi that boasted 345hp, one for each cubic inch. The Chevrolet Corvette and the Chrysler 300-B also had engines that boasted one horsepower per cubic inch in 1957, but these were optional engines, not standard as the Adventurer's was. Unfortunately, this was the last year the Hemi engine was available. Production reached 1,650 for the hardtop and 300 for the convertible.

The 1958 Adventurer was essentially the same car. Again introduced two months later than the rest of the DeSoto line, it featured a minor facelift. However, mechanically, the Hemi was replaced by a two four-barrel 350ci wedge, the Turboflash V-8, rated at 345hp. Low-end performance was similar to the Hemi's but midrange and upper rpm perfor-

1958 DeSoto Adventurer

Engine	
Type	V-8
Bore x stroke, in.	4.12 x 3.38
Displacement, cubic in.	361
Compression ratio	10.25:1
Horsepower	345@5000
Torque	400@3600
Chassis and drivetrain	
Transmission	3-spd automatic
Front suspension	Independent, torsion bars
Rear suspension	Live axle, leaf springs
Axle ratio	3.31:1
Brakes, front/rear	Drum/drum
General	
Wheelbase, in.	126
Height, in.	55
Width, in.	78.2
Length, in.	218.6
Weight, lb.	4,000

All 1959 Adventurers came with front swivel seats. *Galen Erb*

mance was noticeably poorer. A 355hp version, however, was available as an option. This engine featured the Bendix fuel injection system that was tried on other Chrysler engines, with the same poor results. All were recalled to have the unreliable fuel injection replaced with carburetors.

Dual 4-barrel 383 wedge was standard equipment. *Galen Erb*

1959 DeSoto Adventurer	
Engine	
Type	V-8
Bore x stroke, in.	4.25 x 3.38
Displacement, cubic in.	383
Compression ratio	10.1:1
Horsepower	350@5000
Torque	425@3600
Chassis and drivetrain	
Transmission	3-spd automatic
Front suspension	Independent, torsion bars
Rear suspension	Live axle, leaf springs
Axle ratio	3.31:1
Brakes, front/rear	Drum/drum
General	
Wheelbase, in.	126
Height, in.	55
Width, in.	78.7
Length, in.	221.1
Weight, lb.	3,980

The recession of 1958 hurt DeSoto badly; sales hit a low of 49,445 units, which was 13,000 units less than Ford's jumbo failure, the Edsel. Adventurer production in 1958 was 350 hardtops and eighty-two convertibles.

Things continued to slide for DeSoto in 1959. Production was consolidated within the Chrysler Division's Jefferson Avenue plant, while DeSoto's plant at Warren, Michigan, was assigned to build Imperials. Sales, unfortunately, continued to decline to new lows.

Still, the 1959 Adventurer was a fine car. Color choice was either white or black, and the car shared in all the innovations of other 1959 Chrysler products: swivel front seats, automatic headlight dimmer, and so on. The basic bodyshell was the same one as introduced in 1957. The Turboflash V-8 was enlarged to 383ci and 350hp, again with two four-barrel carburetors. Production of the Adventurer hardtop was only 590 units and only ninety-seven convertibles.

The year 1960 marked the last Adventurer. Now built as a unibody, it was available as a two-door hardtop, four-door sedan and four-door hardtop—and unfortunately the make lost the exclusivity it had enjoyed in the past. On the whole, the DeSoto line suffered degradation because only one other model besides the Adventurer was offered. Since it looked like a Chrysler and cost almost as much, it helped to decide the DeSoto's fate, at

The 1960 Adventurer was no longer the high-performance specialty automobile it once was, yet is was still distinctive. This is Alan Orenstein's pristine restoration.

The swivel seats never really caught on.

1960 DeSoto Adventurer

Engine	
Type	V-8
Bore x stroke, in.	4.25 x 3.38
Displacement, cubic in.	383
Compression ratio	10.1:1
Horsepower	305@4600
Torque	410@2400
Optional engine	383-4bbl, 383 Ramcharger
Chassis and drivetrain	
Transmission	3-spd automatic
Front suspension	Independent, torsion bars
Rear suspension	Live axle, leaf springs
Axle ratio	2.93 or 3.31:1
Brakes, front/rear	Drum/drum
General	
Wheelbase, in.	122
Height, in.	54.8
Width, in.	79.4
Length, in.	217
Weight, lb.	3,945

least from the buying public's point of view. The 1961 DeSoto with its unusual two-tier grille nailed the coffin shut. There were plans for a 1962, but the prototype looked as if it had been styled by someone who had seen too many horror movies.

Prospects

The DeSoto Adventurer right now is in the sleeper category, as is the Pace car. From a historical perspective, these were the most luxurious and expensive, fastest and rarest of all DeSotos. Interestingly, these cars also have a poor survival rate, so few exist.

Chapter 4

1967–74 Plymouth Belvedere GTX and Road Runner

★★★	**1967–71 Plymouth Belvedere/Satellite, GTX**
★★★	**1968–74 Plymouth Road Runner**
★★★★	**1970 Plymouth Super Bird**
★★★★	**With 440 6-bbl V-8**
★★★★★	**With 426 Hemi V-8**
★★★★	**Convertibles**

,When one thinks of a Chrysler muscle car the picture that comes to mind is a big, rumbling, intermediate, whether a GTX, a Charger, a Super Bee, or a Road Runer. These were tough street terrors that relied on simplicity (as far as styling goes) and cubic inches to establish their supremacy. Sure GM and Ford had their hot runners, but it was the Chrysler supercars that ran well and reliably, even with the four-barrel 383ci base engine. To get a comparable Ford to perform you had to buy the top engine option, usually a 428CJ, and then "play" with it to get it to run. GM had the very popular GTO, but how secure would you feel with an engine that came with a cast-iron crank and cast rods and pistons? The big Chrysler cars ran well "out of the box"—you didn't have to do anything to them, and if you wanted massive overkill there was always the awesome 426 Hemi V-8. The Hemi-powered cars offered incredible acceleration—no other American production engine could match it. Chevrolet enthusiasts may contend that the L-88-powered Corvette was a match, but it was barely streetable, and to discourage non-competition use, L-88 powered cars did not even have a heater! Chrysler was committed to performance you could use; from 1966 to 1971 about 11,000 Hemi-powered cars were built. No other manufacturer came close to producing that many street cars with what was essentially a slightly detuned race engine.

Chrysler first used the 426 Hemi during the 1964 Daytona 500 stock car race. Although the 413ci and 426ci Max Wedge engines were doing well in drag racing, they were not as successful on stock car race tracks. In 1961 and 1962, Pontiacs won the most races, and in 1963, Ford took the lead. Chrysler needed a strong race engine to dominate the NASCAR tracks which would in turn help Chrysler's performance image in the marketplace.

Chrysler had had experience with hemi-head engines since 1951 when the 331ci Fire-Power V-8 was introduced. The ultimate development of that engine was reached in 1957 with a displacement of 392ci. The FirePower-equipped Chrysler 300s easily swept the NASCAR circuit, but by 1958, Chrysler had stopped producing the "early hemi," as it is now commonly called, for cost reasons. Because there wasn't enough time to develop a new engine from the ground up, Chrysler basically applied the early Hemi's technology to the RB (Raised Block) 426 block. The stress-relieved block featured cross-bolted main bearing caps, and to ensure durability, every component was designed, built, and tested for use only in the 426 Hemi.

The hemi-powered Plymouths easily won the 1964 Daytona 500, finishing 1-2-3, and managed to win twenty-six out of the sixty-two Grand National NASCAR races that year. Still, Ford managed to win thirty races because of better preparation and durability. Rule disputes kept the Hemi-powered cars out of racing for most of 1965. Rules for the

At a quick glance, the 1967 Belvedere GTX looks like a typical sixties family sedan. This example is equipped with the 426 Hemi.

1966 season made it very difficult for Chrysler and for Ford's SOHC 427, both limited production engines, to race unless they were installed in regular production cars. Chrysler felt that the advantages of winning on the racetrack outweighed the cost of producing street versions of the Hemi, thus beginning in 1966, the street Hemi was available in Dodge and Plymouth cars.

The street Hemi did not differ drastically from the race version. It used a milder solid lifter camshaft; lower, 10.25:1, compression ratio; an aluminum, inline, dual plane intake manifold with two Carter four-barrel carbs; and cast-iron exhaust manifolds. It was rated at 425hp at 5000rpm with 490lb-ft torque at 4000rpm. These specifications remained unchanged until 1971, the last year the engine was available, even though camshaft specs were changed in 1968 (slightly longer duration and overlap). Hydraulic lifters were used on 1970 and 1971 engines.

Production stopped in 1971 for the well-known reason that high insurance premiums made the engine too costly, but NASCAR had also changed the rules once again, this time limiting engine size to 305ci.

Obviously, the Hemi-powered 1966–71 intermediates using the B-body platform and the 1970–71 E-body pony cars—the Barracuda and Challenger—are the most desirable of all Chrysler-built cars of the 1960's and early 1970's. It is critical that you know how to identify one of these monsters. Here are some important things to know and look for.

All Hemi cars used a convertible body with a roof section attached to it. The reason for this is the convertible body is stronger than the regular hardtop. The convertibles also have a torque boxes that connect the rear bulkhead into the side rails in front of each rear leaf spring. This not only stiffens the chassis but also serves as a means of identifying a Hemi car. These boxes can be seen if you take a look underneath. All four-speed cars came with the Dana 60 rearend and none were available with air conditioning.

All Hemi cars came with a thick steel plate welded to the floor of the car above the rear pinion snubber, and all came with a 3/8in I.D. fuel line (as did 440-powered cars). Standard fuel line I.D. was 5/16in.

Hemi cars also have a different K-Frame which is used to mount the engine. They all came with the large radiator support and the large radiator used in air conditioned cars. All automatic-equipped Hemis came with an auxiliary transmission oil cooler, and all cars (with the exception of the 1971 B-bodies) came with a side-mounted windshield wiper motor to provide clearance for the air cleaner or fresh air scoop. All 1969 and later Hemi cars (excepting the SuperBird, Charger Daytona, and 1970 Charger) came with outside air induction as standard equipment. With the exception of factory racers, Hemi-powered cars built in 1968 or earlier did not have standard outside air induction.

1967

The Belvedere and its GTX- and Road Runner-based offshoots were Plymouth's versions of the performance intermediates—equivalent to the Dodge Coronet R/T and Super Bee. The models from the two companies were very similar in styling, but Plymouth outsold their Dodge counterparts by a large margin.

Optional on the 1967 GTX were these hood stripes.

The GTX came with somewhat more visible side stripes in 1968. A new front grille and taillight treatment gave the car a cleaner look. Side-marker lights were government mandated.

This is a rare 1968 GTX Convertible.

The 426ci Hemi V-8 engine was first available in 1966 on the Belvedere and the more deluxe Satellite, making these particular models more desirable than any other 1966 Plymouth. Powerful and expensive, not very many of these were sold.

The performance version of the Belvedere, the GTX, was introduced in 1967, with its standard 375hp 440ci Super Commando engine. Standard equipment included bucket seats, console with floor shifter, heavy-duty suspension, larger drum brakes, the three-speed Torqueflight automatic, simulated hood scoops, chromed exhaust tips, and a pit-stop gas cap on the left rear fender. Optional were power front disc brakes, four-speed manual transmission, and, of course, the 426ci Hemi V-8. The GTX provided a performance image, but it was based on the upscale version of the line and sales were limited.

1968

The 1968 GTX received a minor restyle, with a new front grille and rear taillight treatment providing a cleaner look. New for 1968 were government mandated side-marker light, and also a new side stripe treatment. All GTXs came with bright rocker panel stripes. The GTX came with side-facing, non-functional hood scoops. During the model year, a functional system, called the Air Grabber, became an option on the GTX and Road Runner.

Plymouth enthusiasts were excited about the 1968 introduction of the Road Runner. The Road Runner, which used the Road Runner cartoon character as its moniker, was an el cheapo package designed to crush the competition on the street while being affordable. With a price of $2,869, it was a great straight-line performance car with no frills.

1968 GTX interior. The standard tachometer is on the right side of the instrument panel.

The 440 6-barrel Road Runner was a mid-1969 introduction. Fiberglass hood and black painted steel wheels were standard equipment.

The 6-barrel 440 without the air cleaner—an impressive sight. The system was designed to provide "economical" performance around town as the engine ran only on the center carburetor. The outboard units would come in only when the accelerator pedal was floored.

The Road Runner was based on the bottom-of-the-line Belvedere coupe with swingout rear windows. A midyear introduction was a hardtop which also came with a higher level trim than the coupe. A Custom Decor Package was available.

Even with the standard 335hp 383ci V-8, the Road Runner provided mid-14sec (or better) quarter mile times. It was a single purpose machine and well equipped for the job it was designed to do, with heavy duty suspension, bigger brakes, and stronger driveline than the regular Belvederes. A four-speed manual was included with the 383ci V-8, while the Torqueflite automatic was optional. Other options included power front disc brakes with optional 15in wheels. Of course, the 426ci Hemi was optional, but rarely ordered because of its high cost.

The Road Runner proved to be a marketing success, with over 44,000 units sold.

1969

1969 proved to be even better for Plymouth, with over 80,000 Road Runners sold and 15,000 GTXs. Plymouth saw fit to add a convertible to the Road Runner lineup, and by midyear the 440ci 6-barrel engine, (called Six Pack on Dodges) which bridged the gap between the 426ci Hemi and the base 383ci V-8, had also been added. The 440 6-barrel used three 2-barrel Holley carburetors mounted on an Edelbrock manifold, flowing a total of 1350cfm. Impressive to behold, all that CFM was a little too much for a street engine. The 440 6-barrel-equipped Road Runner was a complete package, from the wild, black fiberglass hood and special (Hemi) suspension and driveline. Its 4.10:1 Dana rear ensured that it was strictly a quarter mile car. In fact,

The Coyote versus Road Runner theme extended to Road Runner air cleaner graphics.

Most high performance convertibles command higher prices than their hardtop counterparts. This is a 1969 426 Hemi-equipped GTX.

The Hemi-powered cars may be expensive to obtain, but the more common 383ci-powered Road Runners are more plentiful and therefore less expensive. This particular example has the rarely seen optional 15in chrome Magnum wheels.

that's where the emphasis of Chrysler's muscle cars was. They were not the best-handling cars, partly due to the late-sixties tire technology and mostly because Chrysler focused on drag and NASCAR racing.

1970

The 1970 GTX and Road Runner were based on the restyled Belvedere and bore an evolutionary resemblance to their predecessors, unlike the Coronet R/T and Super Bee. At first glance they did not seem too different from the 1969 models, yet aside from the roof, these were new cars. Most noticeable was the new front grille, giving the car a more attractive and aggressive look. Other styling features included the simulated rear fender brake scoops, hood bulge, and new taillight panel treatment. The convertible was dropped from the GTX line but was still available on the Road Runner. To counter Dodge's Scat Pack, Plymouth formed the Rapid Transit System which grouped Plymouth's performance cars and enabled them to be marketed as a group, showing that performance was available at all levels.

The GTX's engine and transmission options remained unchanged, with the exception of the addition of the 440 6-barrel as an option. Unlike the 1969 version, the 6-barrel engine did not come with a fiberglass hood. Fresh air could be had with the Air Grabber intake system, optional on either 440ci V-8, but standard on the 426ci Hemi V-8.

A nice option which improved the GTX and Road Runner's appearance and performance was the availability of 15x7in Rallye wheels with F60x15in Goodyear tires. There is

Plymouth's answer to the 1969 Daytona Charger was the 1970 Road Runner SuperBird. The two were visually similar, but the Road Runner used a different nose cone.

Even the wing on the 1970 SuperBird was different than the Daytona's. All SuperBird's came with vinyl roofs.

nothing like a nice set of wheels and tires to enhance a car's performance image, and you'll find many cars today with a set of owner installed Rallye wheels. The Road Runner was still available as a 2-door coupe, hardtop, or convertible. Standard engine was the popular 383ci V-8 which was still rated at 335hp, although compression had dropped to 9.5:1. For 1970, a Holley 4-barrel carburetor replaced the Carter as well. In a cost-cutting move, the standard transmission was a three-speed manual, but the four-speed manual and Torqueflite automatic were optional.

The 440 6-barrel and 426 Hemi were optional, but as in previous years, only a small percentage of cars were equipped with these engines.

One interesting change on the 426 Hemi was the switch to a hydraulic cam for emission reasons. Still rated at 425hp, there was a subtle difference in the way the engine ran as compared to the solid lifter version, with the previous version feeling stronger.

The most interesting Road Runner in 1970 was the SuperBird, Plymouth's NASCAR homologation special. Like the Charger Daytona before it, Plymouth had to build a certain number of these cars if they were to be raced on the NASCAR tracks. Unlike the Daytona where only 500 were required, NASCAR changed the rules for 1970 thus necessitating that at least 2,000 SuperBirds had to be built.

The Dodge Charger had been Chrysler's mainstay in NASCAR racing in the late sixties

The 1970 GTX and Road Runner both received a new front grille, rearend, and side treatment. The side scoop is non-functional. The hood paint stripe on this example is non-stock.

1970 440 6-barrel Road Runner Convertible with the functional Air Grabber hood.

because of its inherently better aerodynamics. Plymouth had been offered its own version of Daytona-type aerodynamics, but it took the defection of Richard Petty to Ford to convince Plymouth that in order to win it would need a similar car. Basically, the Daytona's nose cone and rear mounted wing were adapted to the Road Runner, but there were differences. The SuperBird had to use Dodge Coronet fenders which were more downsloping and a special plug was also needed to correct a hood to nose cone mismatch. You'll also note that the SuperBird had a larger air inlet, and the rear wing was larger and swept back a bit more. Engine availability was the same as the GTX's as was transmission choice. Power steering, power disc brakes, and 14in Rallye wheels were all standard equipment. Larger 15in Rallye wheels were optional.

The SuperBird was available in only seven colors: Alpine white, Lime Light, Lemon Twist, Blue Fire Metallic, Tor-Red, Vitamin C Orange, and Corporation Blue. Black or white vinyl graced the interior, and, reflecting the Road Runner's no-frills heritage, a bench seat was standard, though bucket seats were available.

The SuperBird was an incredible car in 1970 and even more so today, especially when you consider its sheer size. Coupled with the 426 Hemi V-8 (135 built) and depending on axle ratio, 160mph and more is possible.

1971–74

From a performance standpoint, 1971 proved to be the last hurrah for the GTX and Road Runner. It was the last year for the GTX which continued to be available with the

The Road Runner and GTX were restyled for 1971. Rounded quarter panels and fenders gave the car a more muscular look. This is a GTX.

The 1971 GTX from the rear.

To help the performance car enthusiast battle ever-rising insurance premiums, the 275hp 340ci V-8 was added to the option list on the 1971 Road Runner. The 340 was installed in 1,681 Road Runners.

Production of the 426 Hemi Road Runner amounted to a mere fifty-five units in 1971.

same engine lineup as before. Engine output was marginally less due to a slight compression reduction, and of course, a lot less when measured under the new net horsepower system.

The Satellite body in GTX and Road Runner trim looked more aggressive and muscular than ever. The "fuselage" look which made Chrysler's big cars look really big and heavy, looked just right in the intermediates.

The Road Runner was offered to the public with the trusty 4-barrel 383ci V-8 rated at 300hp. The big 440 6-barrel and 426ci Hemi V-8 were still optional, but in an effort to help the younger buyer with the ever increasing insurance rates, the small block 340ci V-8 rated at 275hp was made an option at midyear. It didn't help much, as sales continued to slide. Interesting options for 1971 included a rear deck spoiler and rear window louvers much like the ones available on the Barracuda/Challenger.

More of the same was the order for 1972. The standard engine was now a 400ci V-8 while a 340ci small block V-8 and a 280hp 440ci V-8 were optional. The grille was slightly restyled and revised tape stripes were available. All in all, even though the big guns were gone, at least the basic pieces were still there for the hot rodder to improve upon.

A further departure from the Road Runner's original concept occurred in 1973 and 1974 as the standard engine was a 2-barrel 318ci V-8 rated at 170hp. Still, the 340ci (360ci

Although the Road Runner name would continue to be used, the 1974 Road Runner is considered to be the last of the original Road Runners.

Top engine option on the 1974 Road Runner was the 360ci small-block V-8.

in 1974), 400ci, and 440ci V-8 engines were available. The last true Road Runners featured a restyled nose, as Chrysler backed away from loop bumpers, and the usual revised tape stripe treatments to add some pizzazz in 1974. The 1975 Road Runner was based on the "The Small Fury" and in 1976 there was the Road Runner Package and Road Runner Decor Group for the Volaré, which continued until 1980.

Plymouth's muscle cars are still popular. Although you'll find that the rarer cars, like those equipped with the 426 Hemi, or the SuperBird, tend to be restored to strictly original specifications, many Road Runners are drivers rather than strictly show cars.

Chapter 5

★★★	1967-70 Dodge Coronet R/T
★★★	1968-70 Dodge Super Bee
★★★★	Convertibles
★★★★	With 440 Six Pack V-8
★★★★★	With 426 Hemi V-8

1967–70 Dodge Coronet R/T and Super Bee

Although the Coronet was first introduced in 1965 as Dodge Division's entry in the intermediate field, it was successfully restyled in 1966. Of particular interest to the enthusiast was the availability of the 426 Street Hemi in the two-door model. According to Chrysler expert Galen Govier, at least two four-door Hemi sedans are known to exist. You really had to look hard to tell whether a Coronet came with the Hemi or not. The car was fast but lacked any performance image.

The 1967 Coronet featured a slight grille and taillight change, but of more importance was the introduction of a new performance model. This was the Coronet R/T (Road and Track). Standard engine was the husky 4-barrel 440ci V-8 Magnum, putting out 375hp and 480lb-ft of torque. Standard equipment included bucket seats, console with floor shifter, heavy duty suspension, larger drum brakes, three-speed Torqueflight automatic, and simulated hood scoops. Optional were power front disc brakes, four-speed manual transmission, and, of course, the 426ci Hemi V-8. The R/T grille resembled that of the Charger, but the lights were not retractable. Deleteable side stripes and appropriate R/T identification set the car off.

The 1967 Coronet featured the same frontal treatment as the Charger, but without hidden headlights.

Like the rest of the Coronet line, you could order the R/T with a host of other comfort and trim options. Sales totaled 10,181 units, which included 628 convertibles. Only 283 Hemi-powered R/Ts were built.

1968

The Coronet was restyled for the better in 1968. Mechanically, the R/T remained the same, but to set it off, it could be had with either side stripes or the new bumblebee stripes which wrapped around the tail. The bumble bee stripes became the trademark of the Scat Pack, a grouping of Dodge's high-performance models, which included the Charger R/T, Coronet R/T, and Dart GTS. Around mid-1968, Dodge announced the Super Bee to coincide with Plymouth's Road Runner introduction. The Super Bee was based on the

The 1968 Dodge Coronet R/T. The rear, wrap-around Bumblebee stripes were a Dodge exclusive. The Super Bee used the same bodyshell and was equivalent, but less popular, than Plymouth's Road Runner. *Chrysler Corp*

1968 Coronet R/T.

The 1969 Coronet featured this unusual triple taillight arrangement. *Chrysler Corp*

Coronet 440 two-door coupe and featured pop-out rear windows. As it was intended as an econo street racer, it featured minimal trim but did include the power bulge hood from the Coronet R/T and the instrument panel from the Charger.

Standard engine was the 335hp version of the 383ci V-8 which used many parts from the 375hp 440 engine, such as the cylinder heads with their larger exhaust valves (1.74in versus 1.60in), camshaft, exhaust manifolds, and windage tray. The engine was connected to an efficient 2-1/4in dual exhaust system. Standard transmission was a four-speed manual with the three-speed Torqueflite automatic optional. Drum brakes were standard but power discs were optional. With the disc brakes came 15in wheels and tires rather than the stock 14in rims. The 426ci Hemi V-8 was optional.

1969

Few changes were made to the 1969 models. A single bumblebee stripe was used on both the Super Bee and R/T. Bucket seats became optional for the Super Bee, but bench seats were standard. Other visual changes included optional rear quarter panel scoops and the optional Ramcharger fresh air induction hood which used two functional hood scoops. This hood became standard equipment on 426ci Hemi V-8 R/Ts. Also, the Super Bee was available as a hardtop.

The biggest development in 1969 was the availability of the 440 Six Pack V-8 motor on the Super Bee. This was as close as you could get to Hemi acceleration without spending Hemi money. In fact on the street, the 440 Six Pack could out-accelerate the Hemi to about 70mph or so when the Hemi's superior breathing took over. This was a package which appealed strictly to the street/strip set. Its flat-black fiberglass hood was held in place by four hood pins, and the wheels were plain steel highlighted by five chrome lugs. Goodyear G70x15in Polyglas tires, the only ones available, were hardly sufficient to handle all that torque.

The 440-Six Pack used three 2-barrel Holley carburetors mounted on an Edelbrock manifold, flowing a total of 1350cfm. Though impressive to behold, all that CFM was a bit much for a street engine. The Super Bee came with either a four-speed manual or Torqueflite automatic and used what was the strongest rearend available in a street car, the Dana 60 with 4.10:1 ratio. As usual, drum brakes were standard equipment with front discs optional.

1970

The final year for the Coronet R/T and Super Bee was 1970. The biggest change was the new dual loop grille which to some, had a foreign, not-from-this-planet look to it. The rear quarters were slightly restyled and used non-functional side scoops. Different taillights distinguished the R/T from the Super Bee.

Engine options were expanded for the R/T, the 375hp 440ci V-8 was standard with the 440 Six Pack and 426ci Hemi V-8s optional.

This is the functional hood on the 426-Hemi-equipped 1969 Coronet R/T.

Dummy side scoops were also part of the 1969 Coronet R/T.

The 1970 Coronet R/T's front end treatment was a departure from previous styling with its twin loop bumper. Performance-styling cues included the scooped hood, side scoops and Bumblebee rear stripes.

The 1970 Super Bee.

The 440 Six Pack was available on both the Super Bee and R/T in 1970.

The Six Pack engine used a Chrysler cast-iron manifold rather than the Edelbrock unit to save a few dollars.

The Super Bee was still promoted as a budget supercar with the 383ci V-8 as standard equipment. The 440 Six Pack and 426 Hemi V-8s were optional. Standard transmission was now a three-speed manual, but the four-speed manual and Torqueflite automatic were optional. This year, however, the Six-Pack engine did not come with a fiberglass hood. The Super Bee buyer also had a choice of two stripe treatments, the familiar wrap-around tail stripe or a new C-shaped rear quarter stripe treatment.

Other interesting options included a rear deck spoiler, the Tuff steering wheel, and a hood-mounted tachometer.

Things to Consider

These cars, along with their Plymouth counterparts, represented what the Chrysler supercar of the sixties was all about. The primary focus was on the lower end of the market (in terms of cost), but they sure did go. The 440 Six Pack and 426 Hemi cars are rare and very desirable, especially the convertibles, but the other models can provide an excellent way for the enthusiast to participate in the sixties supercar phenomenon.

Because of the way these cars were built and marketed, it is possible to find Super Bees and R/Ts without power steering, power brakes, air conditioning, or even a radio—just a big engine. You are better off getting the most heavily optioned car you can find which will be more enjoyable to drive and more valuable as well.

Engine and driveline parts are available. As with other unibody cars, avoid rustbuckets. There is also very little doubt that most of these cars were driven very hard, modified, and quite likely abused. Be very careful and be sure to inspect the undercarriage of any car in which you are interested. Bent and welded frames or cracks at the spring mounts are not good signs.

And like any other collectible, you can always end up spending quite a bit more on a restoration than the final product will be.

Chapter 6

★★★	**1966–67 Dodge Charger**
★★★★	**1968–70 Dodge Charger**
★★★	**1971–74 Dodge Charger**
★★★★½	**1969 Dodge Charger 500 and Daytona**
add ★	**w/426 Hemi**

1967–74 Dodge Charger and Super Bee

Although Dodge products had been doing fairly well on the track, it was obvious that if Dodge was to appeal to the ever-more-important youth market, more than winning races was needed. Dodge had a very conservative, blue collar image which just didn't appeal to young people. In 1964, Dodge introduced a "youthfully styled Charger" show car which had some interesting features, but more exciting was the 1965 Charger II show car which greatly resembled the 1966 production Charger. Coupled with a clever advertising campaign—"The Dodge Rebellion wants you!"—Dodge managed to increase sales and change its image.

The 1966–67 Charger featured hidden headlights. A clean design which didn't particularly catch on.

The centerpiece of Dodge's efforts was the Charger, which was based on the Coronet body shell but with a fastback roofline. The front grille resembled the Coronet's, but the headlights were retractable, giving the car a more sporty look.

The Charger came standard with a 2-barrel 318ci V-8 while 2-barrel 361ci and 4-barrel 383ci V-8s were optional. The 318 came with a three-speed manual transmission while the larger V-8s got either a four-speed manual or three-speed automatic. A midyear introduction was the 426 Hemi V-8.

The Charger enjoyed modest success with 37,344 sold, but only 468 were built with the 426 Hemi. It did fairly well on the NASCAR circuit, winning the Manufacturers Championship.

The 1967 version was essentially the same except for minor trim changes. The 361ci V-8 was replaced with a 2-barrel version of the 383ci V-8, and the 375hp 440ci Magnum V-8 joined the option list. The 426 Hemi was still available, but only 118 were produced. With sales down to less than 15,000 units, it was obvious that the fastback Charger was not a big hit in the marketplace.

Chrysler's success on the track, though, was to have a significant effect on styling. The 1966–67 Chargers, successful as they were, just did not look that hot on the street. The 1968 Charger changed all that.

1968

Totally restyled, the Charger was by far Chrysler's best-looking car in 1968 and its best-looking performance car of the sixties. Whereas Chrysler's other muscle cars provided excellent performance, the Charger provided both strong performance and good looks. It was a very smooth, attractive, flowing design.

Model line-up expanded to include the Charger R/T which was similarly equipped as the Dodge Coronet R/T. Standard engine was the 440ci Magnum V-8 rated at 375hp

The rear of the 1966–67 Charger was pure fastback with a full-width taillight treatment.

No need to blare out the fact that this 1967 Charger has the 426 Hemi engine—just small, fender-mounted emblems.

The 1968 Dodge Charger R/T—no doubt the best-looking car to come out of Chrysler's styling studios in the sixties. This particular example has the 426 Hemi engine.

The 1968 Charger is a visually pleasing design from any angle.

with the 425hp 426ci Hemi optional. Standard equipment included the familiar heavy duty suspension and brakes and the three-speed Torqueflite automatic; the four-speed manual was optional. The rear bumblebee stripes were a deleteable option.

The base Charger was available with the 225ci Slant Six or 318ci V-8 as standard equipment with a 2-barrel or 4-barrel 383ci V-8 optional.

The Charger was a success on the street, racking up over 90,000 sales in 1968, but it was not as competitive on the NASCAR circuit because of poor aerodynamics. The recessed grille and rear window caused excessive turbulence. By simply substituting a Coronet grille and installing a custom rear window over the tunnel backlight, Chrysler engineers were able to improve aerodynamics and make the Charger again competitive. However, in order to qualify as a production car, a NASCAR rule stated that at least 500 street units had to be built and sold. These were built by Creative Industries for Dodge.

The 1969 Charger's most noticeable change was the split front grille.

This is the 1969 Charger 500. In order to improve aerodynamics for the NASCAR circuit, the front grille was almost flush with the bumper.

The rear of the 1969 Charger 500 has a fastback roof design—again to improve aerodynamics.

All Charger 500s were equipped with the 426 Hemi or the 440 4-barrel V-8 with either the Torqueflite automatic or four-speed manual transmission. The Charger 500 was a 1969 model, and the first competition versions saw racing action in early 1969. Unfortunately, it proved to be only competitive rather than an overwhelming winner because Ford brought out special aerodynamic versions of their intermediates, namely the Torino Talladega and Cyclone Spoiler.

The 1969 production Charger received redesigned taillights while the grille got a divider. Engines and options remained basically the same, but an SE package was available on both the base and Charger R/T. This was a luxury package which included leather front seats and lots of simulated wood on the dash and steering wheel.

The wildest and most famous Charger was the Daytona. It was an all-out attempt to attain supremacy on the NASCAR tracks. The final product of extensive wind tunnel testing was a car that had no equals on the track and nothing came close to it in terms of visual impact on the street. An 18in long nose cone and high-mounted rear wing are the most apparent changes from standard, yet there are other features which also distinguish it from production Chargers. The rear window treatment was the same as the Charger 500's and the front fenders had reverse facing scoops—necessary on the race versions for tire clearance. The windshield pillars also had special air de-

Because the 1969 Charger 500 wasn't able to do the job on the track, Dodge came out with the radical Daytona Charger in 1969. An extended nose cone and tall rear wing were all that was needed for dominance on the track.

The 1969 Daytona is an impressive sight to behold.

flectors. The Daytona was available with either the 440ci Magnum V-8 rated at 375hp or the 426 Hemi, of which only seventy were built. Total production was 503 units, considerably less than the similar looking 1970 Plymouth SuperBird.

With the addition of the Charger 500 and Daytona, 1969 proved to be the high point for the Charger, at least from a sales point of view.

1970

The 1970 Charger, still based on the same body shell, received a revised front loop bumper which did not detract from its good looks. The R/T got simulated scoops on the doors and instead of the bumblebee stripes, a longitudinal stripe was used. The Charger, though, seems to look best without any stripes. The 440 Six Pack was made available on the Charger R/T, but like the Hemi (232 made), it did not prove very popular, and only 116 were built.

Model lineup for 1970 included a Charger 500, but it should not be confused with the 1969 Charger 500. The 1970 Charger 500 was a dressed-up base model with the 318ci V-8 as standard equipment. The SE package was again available but only with the redesigned (now optional) bucket seats. Interestingly, unlike other Chrysler intermediates, the Charger was not available in 1970 with 15 in wheels.

The Charger for 1970 used the same bodyshell as the 1968–69 versions, but with a chrome front loop bumper.

The Charger received a major restyling for 1971. Although the car is shorter than the 1970, it looks longer. Unique vertical door stripes simulate side vents. Note the front spoiler.

Dodge adopted a Pontiac-like front grille on the 1971 Charger. The 1971s also got hidden windshield wipers.

Even with its new styling, the 1971 Charger looked sleek as ever.

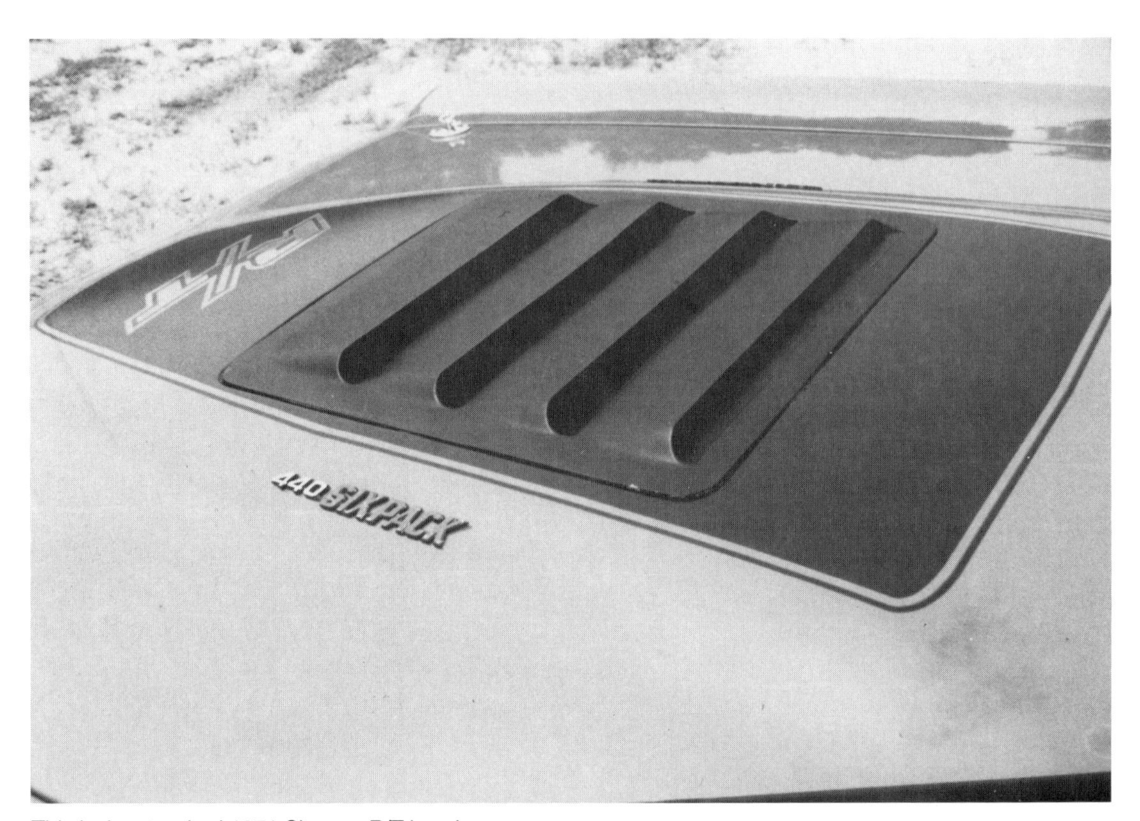

This is the standard 1971 Charger R/T hood.

The Ramcharger hood was optional on the 1971 Charger, but standard with the 426 Hemi engine.

1971–74

The 1968–70 Charger was a hard act to follow, and although the restyled 1971–74 Charger was an attractive automobile, it lacked the one-two punch of its predecessor. The restyled grille had a Pontiac look to it and the high beltline made the car look bigger and heavier even though its wheelbase was 2in shorter and overall length was decreased by 3in.

There were six Charger models for 1971, the base Charger, the Charger hardtop, Charger 500, Charger SE, and more interesting, the Charger Super Bee and Charger R/T.

The Charger Super Bee became Dodge's low buck street racer, replacing the Coronet-based Super Bee of 1970. Standard engine was a 300hp 383ci V-8 with a floor mounted three-speed manual. Optional were the 440ci Six Pack and 426ci Hemi V-8s.

Standard features included Rallye Suspension, heavy duty drum brakes, F70x14in tires, a large Super Bee decal on the blacked-out power bulge of the Ramcharger hood, and a cowl-to-side tape stripe treatment. For the drag racer, the Super Trak Pak Perfor-

The 440 6-barrel V-8 as installed in the 1971 Charger R/T.

The redesigned Charger body of 1971 would continue until the 1974 model year. For 1972, all the high-performance engines were dropped. This is the 1972 Charger SE with its revised side window treatment.

mance Axle Package and 15in Rallye wheels with 60-series tires was available.

Top of the line was the Charger R/T with its standard 440ci Magnum V-8 rated at 370hp. Optional was the 440ci Six Pack and 426 Hemi. The R/T used the same hood and tape stripe treatment as the Super Bee but two additional stripes on each door simulated side vents.

Like the rest of Chrysler's performance offerings, 1971 proved to be the last year for the true high-performance Charger. From 1972–74 the performance model was called the Charger Rallye and was available with increasingly detuned 440 engines.

Things to Consider

Although the 1966–67 Charger is an interesting car, it does not quite have the same appeal as the 1968–70s, though it is one of the more successful full-fastback designs.

The 1968–70 Charger is a classic and a favorite among collectors. The styling has held up and even improved with age. It was one of those designs that stood out, and because of its uniqueness, was not easily copied by other manufacturers. In terms of desirability, the R/T is the obvious favorite, but the base models are also enjoying excellent appreciation based simply on the Charger's good looks.

The 1971–74 Charger lacks the performance image of earlier Chargers, thus their lower prices and slower appreciation. Still a collectible of some stature are the 1971 models because they were still built with all the right engines, but 440 Six Pack and 426 Hemi production was extremely low, less than 400 units.

Chapter 7

1967–74 Plymouth Barracuda

★★	1967–69 Plymouth Barracuda
★★	1970–71 Plymouth Barracuda
★★★	1970–71 'Cuda
★★★★	1970 AAR 'Cuda
★★	1972–74 Plymouth Barracuda
add ★	Convertibles
★★★★	With 440 6-barrel
★★★★★	With 426 Hemi

The Barracuda represented Chrysler's entry in the pony-car market. From a marketing point of view it was never much of a contender, never achieving the acceptance and market penetration that the Mustang, Camaro, or Firebird had. It seemed that Chrysler kept hedging, waiting until 1970 to make a full commitment with a brand new car. In retrospect it

The Barracuda was restyled for 1967 to include this hardtop. This particular Formula S is equipped with the 383ci V-8.

The 1967–69s' hardtop's rearend treatment was somewhat unusual—it was a tad too long.

The 1967 interior was a bit on the spartan side.

was too late; by 1970 the muscle car market was already in decline. On the other hand, it seems that Chrysler stopped production of its pony cars too early. After 1974, the pony-car market saw a resurgence with only GM's Camaro/Firebird to take advantage of it.

The Pony car had its roots with the Chevrolet Corvair introduced by GM in 1960. Designed to appeal to youthful buyers, the Corvair showed that in an industry where car size was equated with profitability (bigger car=bigger profits), a loaded, correctly marketed, compact car could indeed be profitable. Ford rebodied the Falcon, a successful economy compact, called it Mustang and watched sales soar.

Chrysler took a more cautious approach. Still reeling after the styling excesses of the late-Exner era, management concentrated on improving the styling of the larger cars through the efforts of chief stylist Elwood Engel, hired from Ford in 1961. Thus for cost reasons, the first Barracuda, in 1964, was based on the Valiant. The automotive press found the Barracuda to be a very balanced au-

The fastback Barracuda bodystyle was certainly more contemporary, yet it lacked the appeal of a Mustang or a Camaro.

Of the three 1967–69 Barracuda bodystyles, the convertible was most attractive, and when equipped with the 340 small block V-8, a surprisingly quick car.

tomobile exhibiting decent handling with adequate power. It was a good, solid car but didn't have the pizzazz the Mustang had to become a success in the marketplace.

From the perspective of this book, the 1964–66 Barracuda has no collector value. The 1967–69 models will be examined because they are somewhat more desirable as they could be optioned with Chrysler's big-block V-8s. The 1970–74 cars are by far the most interesting and collectible.

1967–69

The first major restyle of the Barracuda occurred in 1967. In addition to the fastback, a convertible and a hardtop were available. The fastback and convertible had likable styling, but the hardtop took some getting used to. Engine choice expanded to include a 383ci 4-barrel V-8 rated at a low 280hp. Because the engine was such a tight fit in the engine compartment, it necessitated the use of restrictive exhaust manifolds, thereby limiting output. Power steering, power brakes, and air conditioning were not available with the big 383, again due to space limitations, making the car a bear to drive.

The performance Barracudas were those equipped with the Formula S package. The Formula S package, available with any 4-barrel engine but only on the fastback body, consisted of heavy duty suspension components, a front anti-sway bar, D70x14in Red Streak tires and front fender Formula S medallions.

The 1970 Barracuda didn't bear any resemblance to the previous cars—making it a true competitor with other pony cars. The high-performance model was the 'Cuda. This is a rare 440 6-barrel convertible.

All 1970 'Cudas got a side "hockey stick" tape treatment which indicated engine size.

This particular 440ci 4-barrel 'Cuda has the larger 15in Rallye wheels. Note the large, Shaker hood scoop.

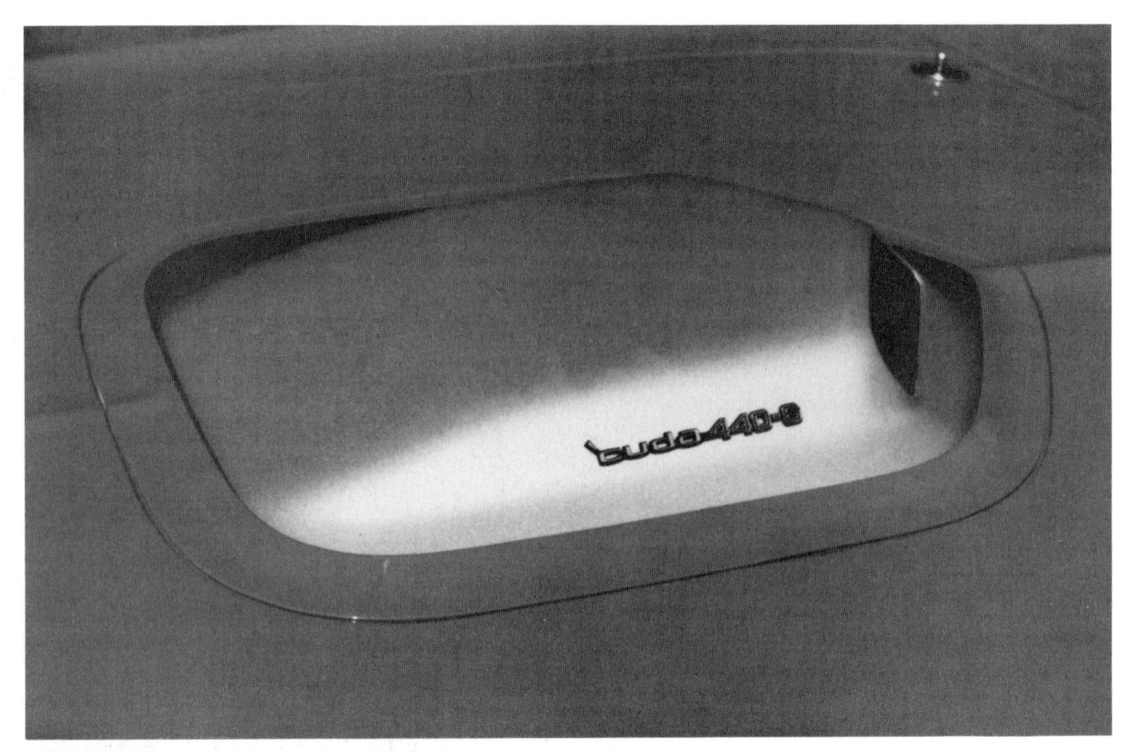

The 'Cuda Shaker was indeed a large piece of fiberglass. It was standard with Hemi-equipped cars and optional on 340-, 383-, and 440-equipped cars.

The ultimate expression of Chrysler power—the 426 Hemi. 'Cudas so equipped were designated "Hemicuda".

Only 1,841 Formula S 383ci V-8-powered Barracudas were built.

The 1968 models benefited from some detail changes. The 273ci V-8 engine was replaced with a 318ci V-8 rated at 230hp, while the base engine for the Formula S, now available on all three body styles, was a 275hp 340ci V-8. The big, heavy 383ci V-8 was still available, now rated at 300hp thanks to revised intake manifolding and cylinder heads. Similar to the 1967 option, it upgraded tire size to E70x14in. For the race track, Chrysler built a limited run of seventy Barracuda S/Ss all painted white and powered by the 426 Hemi. If you are interested in a 1967–69 Barracuda, forget the big blocks and get a 340ci V-8 four-speed. The 340ci V-8 is a good, strong performer that doesn't ruin the car's inherent good balance. The big engines, which include the very limited run of 440ci V-8s of 1969, are just not much fun to drive except in a straight line.

The 1969 Barracuda was a carryover with a new grille. The engine lineup remained the same with the 383ci V-8 now rated at 330hp. A midyear introduction an the hardtop and fastback was the 440ci V-8 rated at 375hp. *Car Life* magazine did a road test on a 440-equipped Barracuda, recording a 0–60mph time of 5.6sec and a quarter mile time of 14.01sec at 103.81mph. The 440ci V-8, though, was only available with the Torqueflite automatic, manual drum brakes, and manual steering.

The "'Cuda" designation was used for the first time in 1969. The 'Cuda 340 and 'Cuda 383 options were available on the hardtop and fastback and consisted of simulated black hood scoops, black hood tape stripes, black lower body tape stripes, 'Cuda 340 or 'Cuda 383 callouts, and chrome exhaust tips. The Formula S was also available, with either the 340 or 383ci V-8s.

The 1967–69 Barracuda proved to be a disappointment to Chrysler. From 1967 to 1969, Ford sold 1,089,000 Mustangs; Chevrolet sold 605,000 Camaros; and Pontiac sold 278,000 Firebirds. Meanwhile, Plymouth managed to move only 139,000 Barracudas. And the competition was getting tougher. The 1969 Mustang could be had with the 428 Cobra Jet engine, the Boss 429 hemi, or the high-winding Boss 302. The Camaro could be had with a hairy 396 or a potent 302 Z-28. And the Firebird could be outfitted with the Ram Air 400.

1970–74

To be competitive, the Barracuda needed a new body that reflected current styling trends and would appeal to the critical youth market—namely a long hood and short deck. The new body also needed to accommodate Chrysler's high-performance engines without compromising the car's balance. These goals were achieved in the 1970 model year when the Barracuda underwent a complete restyle based on the new E-body platform. Also based on the new E-Body was the Dodge Challenger.

The new body style, available only as a notchback or convertible, is considered to be the best looking of the three Barracuda generations, and as we shall see, the most desirable as well.

The 1970 interior provided the driver with more information, especially when equipped with the optional Rallye Instrument Cluster Group.

The wheelbase remained the same at 108in, but overall length was reduced 6in and the height dropped 2in; the width increased by 5in. This made the car look a lot bigger and wider, too wide some said, yet the overall visual effect really depends on what kind of tires the car has. A typical base model with skinny tires does look a bit silly, but the performance models with the 60-series tires look just right.

Model choice expanded to three, the Barracuda, the performance oriented 'Cuda, and the luxury Gran Coupe. All three were available in either the hardtop or convertible. The 'Cuda, to distinguish it from the plain Barracuda, came with a special hood featuring dual nonfunctional scoops, hood pins, driving lights, and a rear taillight panel painted black. Optional was a rear fender "hockey stick" tape stripe which indicated engine displacement. The luxury Gran Coupe came with leather bucket seats (vinyl or vinyl and cloth as reduced cost options) and an overhead console and distinctive taillight trim.

Of course what really made the 1970 Barracudas interesting were the engines. A total of nine engines were available if you include the 198ci Slant Six introduced midyear on the price leader Barracuda Coupe.

A midyear introduction was the wild AAR 'Cuda designed to compete with the Boss 302 and Camaro Z-28- on the street and on the race track. A 340ci V-8 with three 2-barrel carburetor was the only engine available.

Base engine on the Barracuda was a 225ci Slant Six or a 230hp 2-barrel 318ci V-8. Optional were a 2-barrel 383ci V-8 rated at 275hp or a 4-barrel 383ci V-8 rated at 330hp. The Gran Coupe had the same engine availability.

For the performance enthusiasts, the 'Cudas were much more exciting. Standard engine was the 4-barrel 383ci V-8 rated at 335hp. Optional were the 275hp 340ci V-8, 375hp 4-barrel 440ci, 390hp 440-6-barrel V-8, and finally, the mighty 425hp 426ci Hemi V-8. With so many different engines, suspension and tire availability can get confusing.

All 'Cudas came with heavy duty suspensions. The 340 and 383ci V-8 versions came with heavy duty torsion bars, springs, and shocks with front and rear stabilizer bars (0.94in front, 0.75in rear). The big 440ci V-8 and Hemi came with even heavier front torsion bars and unique rear leaf springs, five leafs with two half leafs on the left side and six full leafs on the right side with no rear stabilizer bar. This resulted in a rather firm ride but it was necessary if you wanted to get all that torque to the ground. The 440 and 426ci engines came with a very heavy duty Dana 60 rear axle while the 340/383s came with the Chrysler built 8-3/4in rear in a variety of axle ratios.

Standard brakes were drum front and rear, but these were not really recommended for performance use. The optional power front discs provided better stopping power. Transmission choice was as follows: a three-speed manual was standard with the 340ci and 383ci V-8 engines with a four-speed manual or Torqueflite automatic optional. The 440ci and 426ci V-8 engines came with a specially modified Torqueflite automatic as standard equipment with a four-speed manual optional.

F70x14in tires on Magnum five-spoke rims were standard on 383 or 440ci V-8-equipped 'Cudas. The 340ci V-8 powered 'Cuda came with 15x7in Rallye wheels shod with E60x15in Goodyear Polyglas GTs. The Rallye wheels were also optional on the 383 and 440ci V-8-equipped 'Cudas. F60x15in tires on the 15x7in Rallye wheels were standard on the 426ci Hemi V-8. 'Cuda. The 14in Rallye wheel was optional as well on the 383 and 440's.

Interesting options, some of which were introduced as the year progressed, included the elastomeric color-keyed bumpers. These consisted of unchromed bumpers coated with a painted urethane skin to match the car's paint. These were available in two forms: One option had just the front bumper and racing mirrors colored to match the car, while cars painted in Rallye Red could have both front and rear bumpers painted red with matching racing mirrors. The big shaker hood scoop, standard on the 426ci Hemi V-8, was optional on the other 'Cudas as well. Rear window louvers, front chin spoilers, hood-mounted tach, rim blow steering wheel, "Tuff" sport steering wheel, and a rear deck spoiler were all available to dress up the 'Cuda. Two types of rear deck spoilers were available. One had swept back sides while the other resembled the unit available on Boss 302 Mustangs. In fact it is identical to the Ford unit, as Chrysler purchased it from the same vendor.

The 1970 model year reached its zenith with the introduction of the AAR 'Cuda, which coincided with Plymouth's entry in the popular Trans-Am race series. To qualify, at least 2,500 street versions of the car had to be built. While the full race versions were built by Dan Gurney's All American Racers, the unique fiberglass hood and rear spoiler were built by Creative Industries.

The heart of the AAR 'Cuda was a specially modified 340ci V-8. It used three 2-barrel Holley carburetors mounted on an aluminum Edelbrock intake manifold, special cylinder heads, a solid lifter cam, and a reinforced 340 block. It was rated at 290hp, just like the Boss 302 and Z-28, and provided excellent performance. Car & Driver found it could go from 0–60 in just 5.8sec and the quarter mile came in at 14.3 seconds at 99.5mph.

The rest of the AAR 'Cuda included heavy duty suspension with a rear stabilizer bar, front disc/rear drum brakes, fast-ratio steering, a side-exit exhaust system, and wild side "strobe stripes." The 'Cuda was the only car (besides the Challenger T/A) that came with two different size tires, G60x15in on the rear and E60x15in on the front. A four-speed manual or Torqueflite automatic were the only two transmission choices.

The AAR 'Cuda came with this unique side strobe tape treatment, side exhaust pipes exiting in front of the rear wheels, and a rear deck lid spoiler. Rallye wheels were also standard.

The 1971 Barracuda and 'Cuda got a new front end treatment with four headlights and the infamous "cheese grater" grille. *Richard Steel*

Lacking sufficient development time, the AAR 'Cuda was not successful on the track; however, it was very successful on the street.

Production of 440ci 3x2-barrel-equipped 'Cuda models amounted to 1,755 hardtops and twenty-nine convertibles.

Production of 426ci Hemi-equipped 'Cuda models was 652 hardtops (368 with the automatic and 284 with the four-speed manual) and fourteen convertibles (nine automatics and five manuals).

1971

1971 brought about a restyled grille, commonly known as the "cheese grater" with quad headlights and revised taillights. Other changes included simulated louvers on the front fenders and, along with the color-keyed elastomeric bumpers, a color-keyed grille. Rear side graphics indicating engine displacement replaced the hockey stick stripes of 1970.

Besides the cheese grater grille, the 1971 'Cuda also got front fender air extractors.

In the interior, a split bench seat was optional on the Barracuda and 'Cuda. Leather buckets were still standard on the Gran Coupe and now were optional on the 'Cuda.

The 4-barrel 440ci V-8 was dropped, otherwise engine selection remained the same. Horsepower ratings were lower, as the industry adopted more realistic net horsepower ratings.

'Cuda 440ci 6-barrel production was 237 hardtops and seventeen convertibles. Hemi 'Cuda production was only 108 hardtops (forty-eight with the automatic and sixty with the four-speed manual) and seven convertibles (five automatics and two manuals).

1972–74

Chrysler made a complete withdrawal from the performance market in 1972. All the big-block V-8 engines were dropped—no more shaker hoods, elastomeric bumpers, rear louvers, convertibles, 15in wheels, or 60-series tires. The Barracuda was scheduled for a major restyle, but sales were simply not strong enough to justify the tooling investment. Consequently, the 1972s reverted to a modified single headlight grille, similar to the 1970 but with a divider. Model choice was just two, the Barracuda or the 'Cuda. Standard engines were the 225ci Slant Six or a 2-barrel 318ci V-8. Optional on the Barracuda and standard on the 'Cuda was a 240hp 340ci V-8. No other engines were available. The 'Cuda got side stripes, and for that performance look, you could get a blacked-out hood.

A total of 490 Barracudas were built with the 340ci V-8 along with a total of 5,864 'Cuda 340s.

Little changed for 1973 and 1974. The six-cylinder engine was dropped, and for the limited run 1974 model, the 340ci V-8 engine

Taillight treatment on the 1971s was also revised using separate brake/taillight and backup light housings.

Replacing the hockey stick side stripes on the 1971 'Cuda were these large "billboard" stripes which indicated engine size.

was replaced with a 360ci V-8, which was easier to certify for emission standards. In 1973, 6,583 'Cuda 340s were built.

Things to Consider

From a collector's point of view, the 1970 and 1971 models are the best ones to get. Not many 426 Hemi or 440 6-barrel 'Cudas were built, and these are commanding high prices; natuRallye, the Hemi-powered cars are most desirable. Expect continued appreciation for others with 440 six-pack cars taking the lead. The same applies with the convertibles and the AAR 'Cuda even though the market in general has declined from the highs reached in the late eighties.

Appreciation is more modest with most 340/383/440-powered cars, yet there are some distinct advantages to owning one. They are much less expensive to acquire, and their performance is almost the same—if you are into driving.

The Gran Coupe model got this distinctive emblem.

1971 interior.

This 1972 'Cuda illustrates the 1972–74 taillight and side treatment.

For 1972–74, the Barracuda reverted to a single headlight front end with a divided grille. The 'Cuda came with this type of hood.

In general, as with other Chrysler collectibles, it can be harder and more challenging to restore a Barracuda because there isn't a large cottage industry making reproduction parts. It seems that you can practically build a Mustang or Corvette from the ground up with reproduction parts.

If you are interested in making modifications, try to stay with bolt-ons that are easily reversible, otherwise you're bound to decrease your car's value. When it comes to selling your car, originality is the most important criteria.

Chapter 8

1970–74 Dodge Challenger

★★	1970–1971 Dodge Challenger
★★	1970–1971 Dodge Challenger SE
★★★	1970–1971 Dodge Challenger R/T, RT/SE
★★★★	1970 Dodge Challenger T/A
★★	1972–1974 Dodge Challenger & Rallye
add ★	Convertibles
★★★★	With 440 Six Pack
★★★★★	With 426 Hemi

The Challenger was Dodge's tardy entry in the pony car market. Introduced at the same time as the restyled Plymouth Barracuda of 1970, at first glance, both cars looked as though they shared many body parts. However, the only common part was the window

The 1970 Challenger was Dodge's first pony car. Although similar in size to Plymouth's Barracuda, the Challenger's wheelbase was 2in longer. This is an R/T convertible.

The 1970 R/T version could be equipped with these side stripes. This particular example has had its Magnum wheels widened.

Besides the side stripes, the Challenger R/T could have also been ordered with a rear Bumblebee stripe arrangement. It was an either/or option.

glass. The Challenger was built on a longer 110in wheelbase and featured a prominent S-bend side crease. Unlike the Barracuda, the Challenger used quad headlights set in a grille vaguely reminiscent of the 1969 Camaro and 1969–70 Shelby Mustang. It was a very handsome car, and although styling is subjective, the public seemed to prefer the Challenger over the Barracuda. The Challenger outsold the Barracuda by a significant margin during 1970–74.

Model lineup and engine availability was very similar to the Barracuda. The base model was the Challenger, available as a hardtop or convertible with either the 225ci Slant Six or a 230hp 2-barrel 318ci V-8 as standard equipment. Optional engines were the 275hp 340ci V-8, 290hp 383ci V-8, and the 330hp 383ci V-8. A three-speed manual was standard behind all these engines with the exception of the 290hp 383ci which came only with the three-speed Torqueflite automatic. A four-speed manual and three-speed automatic were optional. The same engine lineup was available on the SE.

The Challenger SE was similar to the Barracuda Gran Coupe except that the SE had a smaller rear window.

For the performance enthusiast, there was the Challenger R/T. Like the 'Cuda, it came with a performance hood, heavy duty suspension, Rallye Instrument Cluster, and a choice of longitudinal or Bumblebee stripes. Engine availability, suspensions, and wheel variations were identical with the 'Cuda. However, you could order the SE package on the R/T which consisted of the formal roof, vinyl roof covering, overhead console, and genuine leather and vinyl bucket seats.

Dodge's answer to the AAR 'Cuda was the Challenger T/A. Specifications were identical, although the Challenger used a different hood, which also cropped up on some R/Ts

For those desiring maximum thrust, the 426 Hemi was optional on the 1970 Challenger.

1970 Challenger R/T front end treatment incorporated this stripe and hood arrangement.

The Challenger T/A complemented Plymouth's AAR 'Cuda. Mechanically identical, the Challenger used a more prominent hood scoop. Note side-exit exhaust system.

as well. Lacking development, the Challenger was not much of a challenge in the 1970 SCCA Trans-Am Series, managing only a distant fourth. On the street, however, it was a different story. The Challenger T/A could easily hold its own against the Boss 302, Camaro Z-28, and the Firebird Trans Am (which had an engine that displaced seventy more cubic inches than the Challenger!).

A version of the T/A was available to West Coast dealers known as the Western Special. It came with a rear exit exhaust system along with Western Special identification on the trunk lid. Some examples also had a vacuum-operated trunk release.

Another midyear entry was the Challenger Deputy which could be had with the small 198ci version of Chrysler's Slant Six. The rear quarter windows were fixed but the Deputy could also be had with the 225in six and the 2-barrel and 4-barrel 383ci V-8s.

For the collector and enthusiast, it's the big-engined Challengers that are of interest. Production of 440 Six Pack cars was relatively high, with a total of 2,035 built, ninety-nine of which were R/T Convertibles. Hemi V-8 production was much less at 356 units with only nine built as convertibles.

1971

For 1971, the Challenger was restyled, or at least the grille was. The split grille insert was painted silver except on the R/T models which were painted black. Other changes for the R/T included color-keyed bumpers, simu-

The Challenger's rear was accentuated by a deck lid spoiler and prominent gas cap.

lated brake cooling scoops in front of the rear wheels, and a new side tape treatment. The SE package was not available on the R/T and the convertible was now available only on the base Challenger.

Engine lineup remained the same for the Challenger R/T. Like the rest of the industry, power ratings were revised to reflect net horsepower rather than the previous gross ratings. Thus, for example, the 425hp 426ci Hemi was now rated at 350hp. No matter how the Hemi was rated, it still provided incredible performance.

Hemi production was very low with just seventy-one units built. There were also only 250 440 Six-Pack-powered Challengers built. The possibility of obtaining one of these cheaply from someone who knows what they have is nearly nonexistent, but one can always hope.

The rare Western Special Challenger T/A came with a vacuum trunk release mechanism as well as a rear exit exhaust system.

Besides the tires, this 1970 Challenger T/A is mostly stock; the decals peel right off. Surprisingly, this car often beats 5.0-liter Mustangs and late-model Camaros and Firebirds.

The Challenger lost most of its high-performance engine options in its last three years of production, but none of its good looks. This is a 1973, with the strobe side tape stripe treatment.

The 1973–74 Challenger can be identified by its larger bumperettes. This is a 1973. The Rallye model came with this non- functional hood.

The last Challenger was a 1974 model. Only 11,354 were produced. *Chrysler Corp*

1972–74

From 1972 on, like the rest of Chrysler's offerings, performance was downplayed. The Challenger lineup changed with only two models offered, the base Challenger and the Rallye (no more convertibles). The name Rallye has never evoked images of thundering performance. Equipped with the 240hp 340ci V-8 engine, the Rallye's performance was in the okay range—0–60mph in the mid-8sec range with quarter miles times in the low 16sec range when the optional 3.55 rear axle ratio was fitted. The Rallye model came equipped with the performance hood, dual exhausts, simulated air extractors on the front fenders, heavy duty suspension with front and rear stabilizer bars, front disc brakes, and a three-speed manual transmission. A four-speed and Torqueflite automatic were optional.

Still, it was a handsome car. The grille from 1972-on was restyled and remained the same until the Challenger's demise in 1974. Engine availability remained the same for 1973, but the Rallye became an optional package on the base Challenger. Standard engine was the 318ci V-8 with the 340ci V-8 optional. During the abbreviated 1974 model run, the 360ci V-8 replaced the 340 as the top engine.

Things to Consider

The Challenger was a good car. The 1970–71 models are the ones to seek as they offer both great performance and collectibility. The Six Pack 440ci V-8s and the Hemi cars may be getting out of reach but the plentiful 383ci and 440ci-4-barrel-powered Challengers are still out there at more reasonable prices. Many of these are still being modified but as long as long as these modifications are largely reversible, it shouldn't hurt the car's value. The market for a stock car is larger than that for a highly modified drag-type car.

Like the Barracuda, Challenger parts availability is not very good. Some body and interior parts are difficult to locate, but engine and driveline parts are relatively plentiful, as they are largely interchangeable with the B-body cars.

Chapter 9

★★★	1968–69 Dodge Dart GTS
★★★	1970–73 Plymouth Duster
★★★	1971–73 Dodge Demon
★★★★	With 440ci V-8

1968–73 Dodge Dart GTS, Dart Sport, Demon 340, Plymouth Valiant Duster 340

The high-performance competition among Detroit's manufacturers extended to the smaller, utilitarian cars of the sixties as well. Lighter and cheaper, these small (by sixties standards) compacts were designed to be simple utility-type vehicles. Still, in typical Detroit fashion, the addition of a few low cost modifications—stripes, styled wheels, dual exhausts—easily

While it may not have the visual appeal of a Charger, the Dart GTS was an able performer for those on a budget. This is a 1969.

The 340ci small block V-8 as equipped on the 1969 Dart GTS.

Yes, the 440ci V-8 does fit in the 1969 Dart GTS—barely.

altered the image of these cars. And even though they were mostly powered by small-block V-8 engines, their performance was generally comparable to larger engined, heavier intermediate muscle cars.

1968–69 Dodge Dart GTS

The Dart GT first appeared in 1965, but the high-performance GTS model did not appear until 1968. To set it off from other Darts, the GTS came with a power bulge hood and GTS identification on the hood, front fenders, and trunk lid—Bumblebee stripes were optional. The GTS was available only as a two-door hardtop or convertible.

A small number of 1969 Dart GTSs were equipped with the 440ci V-8. This one has owner installed widened rear wheels.

As it was the top-of-the-line Dart, the GTS came with upgraded interior trim and standard bucket seats (bench on the convertible with buckets optional.)

The standard engine was the 340ci small-block with the Torqueflite transmission. The 383ci big-block was optional. The four-speed manual transmission was a no cost option on either engine. Heavy duty torsion bars, springs, shocks, and front sway bar characterized the GTS's suspension. Heavy duty drum brakes were standard with front discs optional. Because of the 383's large size, power steering was not available.

A total of eighty 426ci Hemi-powered Dart S/S models were built for drag racing. A small number of dealer-installed 440ci Darts were also built. These were a product of Mr. Norm's Grand Spaulding Dodge of Chicago, Illinois, and they were known as the Dodge Dart GSS.

Sharing the same body style as the 1968, the 1969 GTS got a new grille and taillight panel. The hood had side-facing simulated scoops which also indicated engine size. The GTS also got a different single-band Bumble-

1971 Duster 340.

bee stripe.

Engine choice expanded with the availability of the 375hp 440ci during the later part of the model year. Production figures are unknown, however thirty survivors are known to exist according to Galen Govier.

1970-73 Plymouth Valiant Duster 340

Sharing the same front styling as other Valiants, the Duster 340 was an economy, compact-sized muscle car. Sharing the same frontal styling as the Valiant sedans, the Duster had a fastback roof design and different taillight treatment. To keep costs down, pop-out rear side windows were used.

The Duster 340 could be identified by its flat-black front grille, longitudinal side stripes, and rear tape treatment. The dash was standard Valiant, however a 150mph speedometer took the place of the stock Valiant unit. An 8000rpm tachometer was optional. Bench seats were standard, buckets optional.

The suspension was heavy duty front torsion bars and rear leaf springs with standard manual front disc brakes. The Duster 340 also came with a front sway bar. E70x14in tires mounted on 14in Rallye wheels were also standard.

The 340ci small block, rated at 275hp was mated to a three-speed manual transmission with a floor-mounted shifter. Optional was a four-speed manual and the Torqueflite automatic.

For 1971, the Duster got a new argent painted grille and a different side stripe treatment that ended with large 340 lettering on the rear quarter panels. Although the engine and transmission options remained the same, the Duster 340 was downgraded to the extent that certain standard features became optional, such as the manual front disc brakes and the Rallye wheels. Tire size remained unchanged at E70x14in.

Performance options included a tachometer, disc brakes, the Rallye wheels, Tuff steering wheel, rear wing spoiler, hood

Dodge's version of the 340 Duster was the Demon 340. This is a 1971.

pins.,and a blacked-out performance hood with large "340" lettering. An interesting and rare option was the manual folding sunroof.

The 1972 Duster 340's changes were mostly under the hood. The 340ci engine was now rated at 240hp. Compression ratio dropped to 8.5:1 and the 340 now used the same cylinder heads as the 360ci V-8 which had smaller intake valves (1.88in).

For 1973, Dusters received new front end styling consisting of a different grille design and stronger bumpers. The taillight treatment was similar to 1972, however the taillight bezels were smaller. There were some new items on the option list. The folding sun roof was replaced by a manual metal sunroof, and the rear defogger heated the rear window electrically. Power disc brakes were now standard equipment on the Duster.

The Duster continued after 1973, but the 340ci V-8 was replaced by the 360ci V-8 for 1974-75. Few were built.

1971–72 Dodge Demon

Dodge's response to the Duster was the Demon. Both cars shared the same body and differed only in front grille and taillight treatments. "Demon" decals appeared on the front fenders and on the taillight panel.

Dodge choose not to use the Demon name for 1973. The car was called the Dodge Dart 340 Sport. In 1974, the 340 engine was dropped and replaced by the 360; accordingly, it was named the Dart 360 Sport. The Dart 360 Sport would continue until mid-1976 when it was replaced by the Dodge Aspen.

The 1972 Demon got a new grille and hood scoop arrangement.

Dodge dropped the Demon name in 1972, replacing it with Sport. The sunroof is very rarely seen today, as few were produced. *Chrysler Corp*

Chapter 10

★★	**Fury S/23**
★★★	**Fury GT**
add ★	**w/440 6-barrel**

1970–71 Plymouth Sports Fury GT and S/23

Besides the Chrysler 300, the other full-size performance car available from Chrysler Corporation in 1970 was the Plymouth Sports Fury GT. It was designed to combine luxury and performance in a big car.

Two versions were made, the S/23 and the GT. The S/23 was designed as an "entry level" performance Sports Fury.

Based on the two-door Fury hardtop, standard engine in the S/23 was the 318ci V-8

The 1970 Plymouth Sports Fury was large and sleek. This is a rare S/23. All S/23s and GTs were two doors and came with hideaway headlights.

As with any sixties to seventies vintage muscle car, the larger the engine, the better. This 1970 S/23 has the big 440ci 4-barrel. The 440 6-barrel engine was listed on the option sheets, but only one is known to exist.

with optional engines including the 2-barrel 383ci, the 4-barrel 383ci V-8, and the 4-barrel Super Commando 440ci V-8 rated at 350hp. Standard on the Fury GT was the 4-barrel 440ci V-8 with only one optional engine, the 440ci V-8 6-barrel engine rated at 390hp. All models came with the Torqueflite automatic transmission.

Both cars had the same side strobe tape treatment and hood runner stripes. The hood incorporated two power bulges. The GT also wore chrome exhaust tips. H70x15in tires on steel road wheels were standard with the 440ci V-8 engines, however, the 318 and 383ci V-8s could also be upgraded to this tire size. The S/23 came standard with a front bench seat while the GT got buckets.

The 440 6-barrel engine was the same engine that first appeared in the 1969 Road Runner and Super Bee. Based on the 375hp Magnum 440, the 390hp rating was achieved through the use of a 3x2-barrel carburetor setup. The only difference from the 1969 version is the substitution of a cast-iron intake manifold rather than the Edelbrock aluminum manifold.

According to Chrysler expert Galen Govier, only one 440 6-barrel Fury GT has been located to date.

The Sport Fury GT continued for one more year, 1971, with just the Super Commando 440ci V-8 rated at 370hp (305hp net) mated to the Torqueflite automatic. Power disc brakes, Rallye Road wheels with H70x15in raised-white-letter tires, heavy duty suspension, hood stripes, and side strobe stripes were all standard. Large "GT" letters were located at the back of the hood. The GT's front fender tops were also adorned with three small metal strips, just behind the hood-mounted turn signal indicators. The GT also received the redesigned front grille and rear taillights common to the Fury line.

Things to Consider

Looking at the production figures, it is obvious that the Fury GT or the S/23 didn't find much of a market. Realistically, you could have optioned out a Charger or Belvedere with most of the same luxo type options. It is, however, an interesting, rare, if minor, collectible.

Production Figures

Production of Chrysler 300 Letter Cars

		Hardtops		Convertibles		
Model year	**Series**	**U.S.**	**Exp.**	**U.S.**	**Exp.**	**Total**
1955	C300	1,692	33*	—	—	1,725
1956	300B	1,060	42*	—	—	1,102
1957	300C	1.737	31	479	5	2,252
1958	300D	588	30	187	4	809
1959	300E	534	16	131	9	690
1960	300F	936	28	240	8	1,212
1961	300G	1,266	14	326	11	1,617
1962	300H	435	—	123	—	558
1963	300J	400	—	—	—	400
1964	300K	3,022	—	625	—	3,647
1965	300L	2,373	32	432	8	2,845

**Includes one chassis only.*
Source: Chrysler Corporation

Production of Plymouth Barracudas

1964	
6 cyl	2,647
8 cyl	20,796
Total	23,443
1965	
6 cyl	18,756
8 cyl	41,601
Total	60,168
1966	
6 cyl	10,645
8 cyl	25,536
Total	36,181

1967	**6 cyl**	**8 cyl**
2dr hardtop	10,483	16,277
convertible	859	3,144
2dr sport coupe	5,603	22,425
Grand total		58,791
1968		
2dr hardtop	7,402	11,155
convertible	551	2,044
2dr sports coupe	3,290	16,055
Grand total		40,497
1969		
2dr hardtop	4,203	7,548
convertible	300	973
2dr sports coupe	2,163	12,205
Grand total		27,392

Production of Plymouth Barracudas

1970	6 cyl	8 cyl
Barracuda		
2dr hardtop	5,668	17,829
convertible	223	1,169
Gran Coupe		
2dr hardtop	210	7,184
convertible	34	518
'Cuda		
2dr hardtop	—	17,242
convertible	—	550
Grand total		50,627
1971		
Barracuda		
2dr hardtop	1,555	6,846
convertible	132	721
Grand Coupe		
2dr hardtop	—	1,298
'Cuda		
2dr hardtop	—	5,314
convertible	293	
Grand total		16,159
1972		
Barracuda		
2dr hardtop	809	8,951
'Cuda	—	6,382
Grand total		16,142
1973		
Barracuda	—	9,976
'Cuda	—	9,305
Grand total		19,281
1974	4,989	
Grand total		4,989

**The 1970 model year included 2,724 Plymouth AAR 'Cudas.*

Production of Dodge Challengers

1970*	6 cyl	8 cyl
Challenger		
2dr hardtop	9,929	39,350
2dr fstbk	350	5,873
convertible	378	2,543
Challenger R/T		
2dr hdtp	—	13,796
2dr fstbk	—	3,753
convertible	—	963
Grand total		76,935
1971		
Challenger		
2dr hdtp	1,672	18,956
convertible	83	1,774

Production of Dodge Challengers

1971	6 cyl	8 cyl
Challenger R/T		
2dr hdtp	—	3,814
Grand total		26,299
1972		
Challenger		
2dr hdtp	842	15,175
Challenger Rallye		
2dr hdtp	—	6,902
Grand total		22,919
1973		
Challenger		
2dr hdtp	—	27,930
Grand total		27,930
1974		
Challenger		
2dr hdtp	—	6,063
Grand total		6,063

**The 1970 model year included 2,399 Challenger T/As.*

Production of Dodge Chargers

1966	6 cyl	8 cyl
2dr	—	37,344
Grand total		37,344
1967		
2dr	—	14,980
Grand total		4,980
1968		
2dr	906	74,019
2dr R/T	—	17,665
Grand total		92,590
1969*		
2dr	542	65,840
2dr R/T	—	19,298
Grand total		85,680
1970		
2dr	211	9,163
2dr 500	—	27,432
2dr R/T	—	9,509
Grand total		46,315
1971		
2dr	116	355
2dr Custom	1,441	40,123
2dr 500	—	10,306
2dr SE	—	14,641
2dr Super Bee	—	4,144
2dr R/T	—	2,659
Grand total		73,785

1972		
2dr	813	5,557

Production of Dodge Chargers

1972	**6 cyl**	**8cyl**
2dr custom	132	41,585
2dr SE	—	20,266
Grand total		68,353
1973		
2dr	931	8,857
2dr Custom	374	40,783
2dr SE	—	57,026
Grand total		107,971
1974		
2dr Custom	1,013	23,703
2dr SE	—	30,957
Grand total		55,673

**The 1969 model year included 503 Daytonas and 500 Charger 500s.*

Production of Plymouth Belvedere GTXs and Road Runners

Plymouth Belvedere GTX

1967		**1968**	
2dr hdtp	11,429	2dr hdtp	17,246
convertible	686	convertible	1,026
Total	12,115	Total	18,272
1969		**1970**	
2dr hdtp	14,385	2dr hdtp	7,202
convertible	625	**1971**	
Total	15,010	2dr hdtp	2,626
Grand total			55,225

Plymouth Road Runner

1968			**1969**
2dr coupe	29,240	2dr coupe	32,717
2dr hdtp	15,358	2dr hdtp	47,365
convertible	2,027		
Total	44,598	Total	82,109
1970		**1971**	
2dr coupe	14,744	2dr hdtp	13,046
convertible	684	**1972**	
2dr hdtp	20,899	2dr hdtp	6,831
Super Bird	2,783*	**1973**	
Total	39,110	2dr hdtp	17,443
1974			
2dr hdtp	9,656		
Grand total			212,793

**Other quoted production figures are 1,920, 1,935 and 1,971.*

Production of Dodge Coronet R/Ts

1967		**1968**	
2dr hdtp	9,553	2dr hdtp	9,989
convertible	628	convertible	569
Total	10,181	Total	10,558
1969		**1970**	
2dr hdtp	6,518	2dr hdtp	2,172
convertible	437	convertible	236
Total	6,955	Total	2,408
Grand total			30,102

Production of Dodge Super Bees

1968		**1969**	
2dr coupe	7,842	2dr coupe	7,650
2dr hdtp			18,475
Total			26,125
1970		**1971**	
2dr coupe	3,640	2dr hdtp	4,144
2dr hdtp	10,614		
Total	14,254		
Grand total			52,365

Production of Plymouth Fury GTs

1970		**1971**	
2dr hdtp (PP-23)	666	2dr hdtp (PP-23)	333
2dr hdtp (PS-23)	689		
Total	1,355		
Grand total			1,688

Production of Plymouth Duster 340s

1970	21,799	**1971**	10,403
1972	14,132	**1973**	13,862
Grand total			60,196

Production of Dodge Demon 340s

1971	7,899	**1972**	8,739
1973	9,880		
Grand Total			26,518

Production of Dodge Dart GTSs

1968		**1969**	
2dr hdtp	7,699	2dr hdtp	5,717
convertible	403	convertible	360
Total	8,102	Total	6,077
Grand total			14,179

Source: Chrysler Corporation.

Production of DeSoto Pacesetter and Adventurers

Pacesetters

1956

convertible 100-400 (est.)

Adventurer

1956		**1957**	
2dr hdtp	996	2dr hdtp	1,650
Total	996	convertible	300
Total			1,950
1958		**1959**	
2dr hdtp	350	2dr hdtp	590
convertible	82	convertible	97
Total	432	Total	687

1960

2dr hdtp	3,092
4dr hdtp	1,958
4dr sedan	9.032
Total	14,082

Grand total Adventurer production 18,147

Production of Plymouth Furys

1956-58

1956		
2dr hdtp	4,485	
1957		
2dr hdtp	7,438	
1958		
2dr hdtp	5,303	
1956-58 Total		17,226
1959		
Plymouth Fury		
2dr hdtp	21,494	
2dr hdtp	13,614	
4dr sedan	30,149	
Plymouth Sport Fury		
2dr hdtp	17,867	
2dr convertible	5,990	
1959 Total		89,114

Serial numbers of 1969-72 440 three two-barrels

Location 1969-72: on a plate attached to the left side of instrument panel, visible through windshield.
First digit: Car line
B = Barracuda
J = Challenger
P = Fury
R = Belvedere/Satellite
W = Coronet 1969-70
W - Charger 1971
X = Charger 1970-71
Second digit: Price class
H = High
M = Medium
P = Premium
S = Special
Third and fourth digits: Body type
21 = 2dr coupe
23 = 2dr hardtop

Serial numbers of 1969-72 440 three two-barrels

27 = convertible
29 = 2dr sports hardtop
Fifth digit: Engine
M = Special-order V-8 engine 1969 A12 pkg
V = 440 3x2bbl engine
Sixth digit: Model year
9 = 1969
0 = 1970
1 = 1971
2 = 1972
Seventh digit: Assembly plant
A = Lynch Road, MI
B = Hamtramck, MI
E = Los Angeles, CA
F = Newark, DE
G = St. Louis, MO
R = Windsor, ONT Canada
Last six digits indicate sequence number.

Production of 1969-72 440 three two-barrels

Vin #	Year	Model	Body style	Total
WM21M9	1969	Super Bee	2dr coupe	1,487
WM23M9	1969	Super Bee	2dr hardtop	420
RM21M9	1969	Road Runner	2dr coupe	615
RM23M9	1969	Road Runner	2dr hardtop	817

Vin #	Year	Model	Body style	Total
JS23V0	1970	Challenger R/T	2dr hardtop	1,640
JS27V0	1970	Challenger R/T	convertible	99
JS29V0	1970	Challenger R/T SE	2dr sports hardtop	296
WM21V0	1970	Super Bee	2dr coupe	196
WM23V0	1970	Super Bee	2dr hardtop	1,072
WS23V0	1970	Coronet R/T	2dr hardtop	194
WS27V0	1970	Coronet R/T	convertible	16
XS29V0	1970	Charger R/T	2dr sports hardtop	116
BS23V0	1970	'Cuda	2dr hardtop	1,755
BS27V0	1970	'Cuda	convertible	29
RM21V0	1970	Road Runner	2dr coupe	651
RM23V0	1970	Road Runner	2dr hardtop	1,846
RM23V0	1970	Super Bird	2dr hardtop	716
RM27V0	1970	Road Runner	convertible	34
RS23V0	1970	GTX	2dr hardtop	678
PP23V0	1970	Sport Fury GT	2dr hardtop	1*
JS23V1	1971	Challenger R/T	2dr hardtop	250
WM23V1	1971	Super Bee	2dr hardtop	99
WS23V1	1971	Charger R/T	2dr hardtop	178
BS23V1	1971	'Cuda	2dr hardtop	237
BS27V1	1971	'Cuda	convertible	17
RM23V1	1971	Road Runner	2dr hardtop	246

Production of 1969-72 440 three two-barrels

Vin#	Year	Model	Body style	Total
RS23V1	1971	GTX	2dr hardtop	135
WH21V2	1972	Charger Rallye	2dr coupe	0*
WH23V2	1972	Charger Rallye	2dr hardtop	2*
RM23V2	1972	Road Runner/GTX	2dr hardtop	1*

**When a production total is not known, records show the number known to exist. If proof of production is known, but none are known to exist, records will show 0. Records are for US specifications.*
Source: Galen Govier

Serial numbers of 1966-71 426 Hemi

Location 1966-67 on a plate attached to the left front hinge pillar post.
Location 1968-71 on a plate attached to the left side of instrument panel,
visible through windshield.
First digit: Car line
B = Barracuda
J = Challenger
L = Dart
R = Belvedere/Satellite
W = Coronet 1969-70
W = Charger 1971
X = Charger 1966-70
Second digit: Price class
E = Economy
L = Low
M = Medium
O = Super stock
P = Premium
S = Special
X = Fast top
Third and fourth digits: Body type
21 = 2dr sedan/coupe
23 = 2dr hardtop
27 = convertible
29 = 2dr sports hardtop
41 = 4dr sedan
Fifth digit: Engine
H = 426 2x4bbl Hemi engine stage I mechanical lifters 5qt oil pan 1966

J = 426 2x4bbl Hemi engine stage I mechanical 5qt oil pan 1967
R = 426 2x4bbl Hemi engine stage II mechanical lifters 6qt oil pan 1968-69
M = Special-order V-8 enigne stage III hydraulic lifters 6qt oil pan 1970-71
Sixth digit: Model year
6 = 1966
7 = 1967
8 = 1968
9 = 1969
0 = 1970

Seventh digit: Assembly plant
1 = Lynch Road, MI 1966-67
A = Lynch Road, MI 1968-71
2 = Hamtramck, MI 1966-67
B = Hamtramck, MI 1968-71
7 = St. Louis, MO 1966-67
G = St. Louis, MO 1968-71
9 = Windsor, ONT Canada 1966-67
R = Windsor, ONT Canada 1968-71
Note: 426 Hemi engine was not installed at Los Angeles assemply plant. Last six digits indicate sequence number.

Production of 1966-71 426 Hemi

Vin#	Year	Model	Body style	Total
WE21H6	1966	Coronet	2dr sedan	34
WE41H6	1966	Coronet Deluxe	4dr sedan	2*
WL21H6	1966	Coronet Deluxe	2dr sedan	49
WH23H6	1966	Coronet 440	2dr hardtop	288
WH27H6	1966	Coronet 440	convertible	6
WH41H6	1966	Coronet 440	4dr sedan	1*
WP23H6	1966	Coronet 500	2dr hardtop	340
WP27H6	1966	Coronet 500	convertible	21
XP29H6	1966	Charger	2dr sports hardtop	468
RL21H6	1966	Belvedere I	2dr sedan	136
RH23H6	1966	Belvedere II	2dr hardtop	531
RH27H6	1966	Belvedere II	convertible	10
RH45H6	1966	Belvedere II	6 p wagon	1*
RP23H6	1966	Satellite	2dr hardtop	817
RP27H6	1966	Satellite	convertible	27
WL21J7	1967	Coronet Deluxe	2dr sedan	2*
WS23J7	1967	Coronet 440	2dr hardtop	1*
WO23J7	1967	Coronet 440 S/S	2dr hardtop	55
WP23J7	1967	Coronet 500	2dr hardtop	0*
WS23J7	1967	Coronet R/T	2dr hardtop	59
WS27J7	1967	Coronet R/T	convertible	2
			Combined total, hardtop and convertible	283
XP29J7	1967	Charger	2dr sports hardtop	118
RH21J7	1967	Belvedere I	2dr sedan	2*
RO23J7	1967	Belvedere II S/S	2dr hardtop	55
RH23J7	1967	Belvedere II	2dr hardtop	6*
RH41J7	1967	Belvedere II	4dr sedan	1*
RP23J7	1967	Satellite	2dr hardtop	6*
RP27J7	1967	Satellite	convertible	2*
RS23J7	1967	Belvedere GTX	2dr hardtop	108
RS27J7	1967	Belvedere GTX	convertible	17
LO23M8	1968	Dart S/S	2dr hardtop	80
WM21J8	1968	Super Bee	2dr coupe	125
WH21J8	1968	Coronet 440	2dr coupe	2*
WS23J8	1968	Coronet R/T	2dr hardtop	220

Vin#	Year	Model	Body style	Total
WS27J8	1968	Coronet R/T	convertible	9
XS29J8	1968	Charger R/T [Includes one Charger 500 prototype]	2dr sports hardtop	475
BO29M8	1968	Barracuda S/S	2dr sports hardtop	70
RM21J8	1968	Road Runner	2dr coupe	840
RM23J8	1968	Road Runner	2dr hardtop	171
RS23J8	1968	GTX	2dr hardtop	410
RS27J8	1968	GTX	convertible	36
WM21J9	1969	Super Bee	2dr coupe	166
WM23J9	1969	Super Bee	2dr hardtop	92
WS23J9	1969	Coronet R/T	2dr hardtop	97
WS27J9	1969	Coronet R/T	convertible	10
XS29J9	1969	Charger R/T	2dr hardtop	432
XX29J9	1969	Charger 500	2dr fast top	67*
XX29J9	1969	Charger Daytona	2dr fast top	70
RM21J9	1969	Road Runner	2dr coupe	356
RM23J9	1969	Road Runner	2dr hardtop	422
RM27J9	1969	Road Runner	convertible	10
RS23J9	1969	GTX	2dr hardtop	198
RS27J9	1969	GTX	convertible	11
JS23R0	1970	Challenger R/T	2dr hardtop	287
JS27R0	1970	Challenger R/T	convertible	9
JS29R0	1970	Challenger RT/SE	2dr sports hardtop	60
WM21R0	1970	Super Bee	2dr coupe	4
WM23R0	1970	Super Bee	2dr hardtop	32
WS23R0	1970	Coronet R/T	2dr hardtop	13
WS27R0	1970	Coronet R/T	convertible	1*
XS29K0	1970	Charger R/T	2dr sports hardtop	112
BS23R0	1970	Hemi 'Cuda	2dr hardtop	652
BS27R0	1970	Hemi 'Cuda	convetible	14
RM21R0	1970	Road Runner	2dr coupe	74
RM23R0	1970	Road Runner	2dr hardtop	75
RM23R0	1970	Super Bird	2dr hardtop	135
RM27R0	1970	Road Runner	convertible	3
RS23R0	1970	GTX	2dr hardtop	72
JS23R1	1971	Challenger R/T	2dr hardtop	71
WM23R1	1971	Super Bee	2dr hardtop	22
WS23R1	1971	Charger R/T	2dr hardtop	63
BS23R1	1971	Hemi 'Cuda	2dr hardtop	108
BS27R1	1971	Hemi "Cuda	convertible	7
RM23R1	1971	Road Runner	2dr hardtop	55
RS23R1	1971	GTX	2dr hardtop	30

**When a production total is not known, records show the number known to exist. If proof of produciton is known, but none are known to exist, records will show 0. Records are for US specifications.*
Source: Galen Govier.

Clubs

Mopar clubs

W.P.C. Club, Inc.
P.O. Box 3504
Kalamazoo, MI 49003

Founded 1969. Publishes *WPC News*, a high-quality monthly publication, featuring all Chrysler eras.

MoPar Muscle Club International
Route 9, Box 18
Lockport, Il 60441

Founded 1978, primarily focuses on 1960s and 1970s high-performance Chrysler cars. Publishes *Muscle Hustle*, an excellent monthly with many interesting features and articles.

Chrysler 300 Club, Inc.
13333 Branchwater Lane
Birmingham, AL 35216

Founded 1969, covers all Chrysler 300s, Letter and non-Letter. Publishes *Brute Force* bimonthly.

Chrysler 300 Club International, Inc.
19 Donegal Court
Ann Arbor, MI 48104

Founded 1969, caters primarily to 300 Letter Series, although ownership is not required.

DeSoto Club of America
105 E 96th
Kansas City, MO 64114

Founded 1972, dedicated to preserving and restoring the DeSoto automobile. Publishes newsletter *DeSoto Days*.

National DeSoto Club, Inc.
412 Cummock Road
Inverness, Il 60067

Founded 1986. Publishes *DeSoto Adventures* bimonthly.

Plymouth Barracuda/'Cuda Owners Club
RD4, Box 61
Borthampton, PA 18067

Founded 1978, devoted to Barracuda/'Cuda. Bimonthly publication.

National Hemi Owners Association
170 Pansy Pike
Blanchester, OH 45107

Founded 1975, open to owners and enthusiasts of Chrysler Hemi-powered vehicles.

Northeast Hemi Owners Association
P.O. Box 426
St. Peters, PA 19470

Dedicated to preserving, restoring, and enjoying Chrysler performance vehicles. Publishes bimonthly newsletter.

Daytona/Super Bird Automobile Club
13717 Green Meadow
New Berlin, WI 53151

Dedicated to the Daytona and Super Bird.

National Chrysler Products Club
P.O. Box 3150
Falls Church, VA 22043

National club covering all Chrysler cars.

Index

clubs, 127
Cunningham, Briggs, 9
"The Dukes of Hazzard", 21
Engel, Elwood, 31
Exner, Virgil, 55
Govier, Galen, 74
Granatelli, Andy, 31
Gurney, Dan, 99
Hemi V-8, 7, 8, 16, 20, 55, 59, 60, 68
Kiekhaefer, Carl, 9, 14
Mr. Norm's Grand Spaulding Dodge of Chicago, 114
Petty, Richard, 69
production figures, 120
Rapid Transit System, 66
Rushing, Jerry, 20, 21
Traveller, 18, 20, 21

Models:
C300, 6–11
300-B, 10, 12–15
300-C, 11–17
300-D, 16, 18, 20
300-E, 21–23
300-F, 24, 26, 28
300-G, 27, 28, 31
300-H, 29–34
300-J, 31, 34–36
300-K, 35, 37–39
300-L, 39
Charger 500, 84, 86
Charger II show car, 80
Daytona, 69, 84, 86
DeSoto Adventurer, 48
Dodge
1970–74 Challenger, 105–112
1967–74 Charger, 80–90
1967–70 Coronet R/T, 74–79
1968–69 Dart GTS, 114
1968–73 Dart GTS, 113
1968–73 Dart Sport, 113
1968–73 Demon 340, 113
1971–72 Demon, 116, 117
1967–74 Super Bee, 74–79
Plymouth,
1967–74 Barracuda, 91–104
1967–74 Belvedere GTX, 59–73
1956–58 Fury, 40
1967–74 Road Runner, 59–73
1970–71 Sports Fury GT, 118
1970–71 Sports Fury S/23, 118
1970–73 Valiant Duster 340, 115, 116
SuperBird, 68, 69